DATE DUE

NOV 1 6 2000	
APR 1 6 2001	

The Four-Color Theorem

Springer
*New York
Berlin
Heidelberg
Barcelona
Budapest
Hong Kong
London
Milan
Paris
Singapore
Tokyo*

Rudolf Fritsch
Gerda Fritsch

The Four-Color Theorem

History, Topological Foundations, and Idea of Proof

Translated by Julie Peschke

With 42 Illustrations

Springer

Rudolf and Gerda Fritsch
Mathematisches Institut
Ludwig-Maximilians-Universität München
Theresienstrasse 39
D-80333 München, Germany

Julie Peschke (Translator)
Department of Mathematical Sciences
University of Alberta
Edmonton, Alberta T6G2G1, Canada

The cover illustration shows a map explaining the failure in Kempe's "proof" of the Four-Color Theorem. See page 176.

Mathematics Subject Classification (1991): 01A05

Library of Congress Cataloging-in-Publication Data
Fritsch, Rudolf, 1939–
  [Vierfarbensatz. English.]
  The four color theorem / Rudolf Fritsch, Gerda Fritsch.
    p. cm.
  Includes bibliographical references (p. –) and index.
  ISBN 0-387-98497-6 (acid free paper)
    1. Four-color problem. I. Fritsch, Gerda. II. Title.
  QA618.12.F7513 1998
  511'.5–dc21   98-11967

Printed on acid-free paper.

Original title: *Der Vierfarbensatz*, by Rudolf and Gerda Fritsch

Production managed by Lesley Poliner; manufacturing supervised by Thomas King.
Photocomposed copy prepared from the translator's LaTeX file.
Printed and bound by Hamilton Publishing Co., Rensselaer, NY.
Printed in the United States of America.

9 8 7 6 5 4 3 2 1

ISBN 0-387-98497-6 Springer-Verlag New York Berlin Heidelberg SPIN 10557643

To
Dorothee, Veronica, and Bernhard

Preface

During the university reform of the 1970s, the classical Faculty of Science of the venerable Ludwig-Maximilians-Universität in Munich was divided into five smaller faculties. One was for mathematics, the others for physics, chemistry and pharmaceutics, biology, and the earth sciences. Nevertheless, in order to maintain an exchange of ideas between the various disciplines and so as not to permit the complete undermining of the original notion of "universitas,"[1] the Carl-Friedrich-von-Siemens Foundation periodically invites the professors from the former Faculty of Science to a luncheon gathering. These are working luncheons during which recent developments in the various disciplines are presented by means of short talks. The motivation for such talks does not come, in the majority of cases, from the respective subject itself, but from another discipline that is loosely affiliated with it.

In this way, the controversy over the modern methods used in the proof of the Four-Color Theorem had also spread to disciplines outside of mathematics. I, as a trained algebraic topologist, was asked to comment on this. Naturally, I was acquainted with the Four-Color

---

[1] A Latin word meaning *the whole of something, a collective entirety.*

Problem but, up to that point, had never intensively studied it. As an outsider,[2] I dove into the material, not so much to achieve any scientific progress with it but to make this already achieved objective more understandable.

My talk on this subject was given in the winter semester of 1987/88, and it generated interest among my colleagues. This brought to mind my primary jurisdiction—the Professorship for Mathematical Education at the University of Munich. I then began to think about how one could make the mathematical workings of the Four-Color Problem more accessible to student and professor alike—many of whom were already fascinated by this famous problem.

This led to a lecture at the 80th gathering of the *Deutscher Verein zur Förderung des mathematischen und naturwissenschaftlichen Unterrichts* (German Association for the Advancement of Teaching in Mathematics and the Sciences), which was held in 1989 in Darmstadt [FRITSCH 1990]. The director of B.I. Wissenschaftsverlag, who attended that meeting, approached me about formulating more precisely my thoughts on this matter—from the point of view of an outsider to other interested outsiders. In other words, I was to put them into book form.

Therefore, this book has been written to explain the Four-Color Theorem to a lay readership. It is for this reason that a chapter on the historical development and the people involved in it has also been included. When my efforts concerning the historical side of things bogged down, I managed to persuade my wife to take on this task. She dedicated herself to it wholeheartedly, for which I am truly grateful.

From the point of view of mathematical research, much has already been written about the Four-Color Theorem—before, during, and after the completion of its proof. There exist a number of first-rate complete presentations of which the interested and professional mathematician can avail him/herself. I will mention just a few of them: for the time "prior," the book of Oysten Ore [ORE 1967]; for the

---

[2]I regard as "insiders" those mathematicians who have devoted themselves to graph theory and combinatorics.

period "during," the book of Thomas Saaty and Paul Kainen [SAATY and KAINEN 1977]; and for "after" the fact, the book of Martin Aigner [AIGNER 1984].

In the task set before me, the following problem arose. The intuitive presentation of maps is topological in nature. The ever-widening rings around the proof of the Four-Color Theorem, however, are fraught with purely combinatorial difficulties. The topological side in most of the presentations is somewhat neglected. Perhaps this is because it is viewed as being trivial or uninteresting. Many of the required theorems are so intuitively clear that one considers them not needing any proof at all. Here is an example:

> In the plane, let B be an arc. Let **x** be an end point of B. Then every point not lying in B can be joined to **x** by an arc that intersects B only at the point **x**.

However, a proof of this claim cannot be found in the truly comprehensive textbook *Lehrbuch der Topologie* [RINOW 1975]. On page 409, the author writes: *As we, in the future, will not be needing this particular theorem on arcwise accessibility, we will omit the somewhat long and drawn-out proof.* Nowadays, the much-discussed subject of fractal geometry has, however, sharpened the eye precisely for such problematic issues.[3] From this, it has become apparent that the transition from topology to combinatorics is possible to present—even in the desired way. It just requires deep results of modern topology. These, however, cannot be fully presented within the limited pages of this little book, meaning that all of the proofs cannot be included as well. Nonetheless, the facts are presented, and references for the more extensive proofs are supplied. The most valuable source of information is Dugundji's book *Topology* [DUGUNDJI 1966]. Even this book does not contain everything that is necessary. One must still draw upon particular techniques that come from *piecewise linear* and *geometric* topology. Most of these can be found in Moise's *Geometric Topology in Dimensions 2 and 3* [MOISE 1977].

---

[3]It has even been suggested to include the notion of "monster curves" in the school curriculum. Monster curves are an interesting topic for this book because they highlight the difficulties inherent in the mathematical definition of maps [NEIDHARDT 1990].

The third part of this book treats the combinatorial methods that constitute the actual content of the Four-Color Theorem. This need not be discussed here in any further detail.

Many colleagues and friends must be thanked for their efforts in supplying both historical and mathematical information to us. They were plied with many questions in conversations and in correspondence. In 1990, at the fiftieth anniversary of the *Deutsche Mathematiker-Vereinigung* (German Mathematical Association) in Bremen, Wolfgang Haken gave a full account of this problem. Jean Mayer sent very helpful letters. By means of "electronic mail," Karl Dürre, Kenneth Appel, and John Koch were exceedingly friendly and patient in answering a continuous stream of inquiries. We are not able to mention here all who have helped us. They simply must excuse the omission.

A final thank you goes to Mrs. Bartels and Mr. Engesser of B.I. Wissenschaftsverlag who have obligingly granted special requests for the provision of facsimiles and illustrations and who have gone to great lengths to procure them—sometimes a very difficult task. A friend and colleague of mine who did a great deal in the way of proofreading has requested to remain anonymous.

Rudolf Fritsch
München, Germany
October 23, 1993

Acknowledgments for the English Edition

Our express thanks to Julie Peschke, who not only excellently performed the difficult task of translation but also, as a mathematician, discovered many (hopefully almost all) misprints, inaccuracies, and even errors in the German edition.

Special thanks also go to Tom von Foerster, of Springer-Verlag, New York who accepted the book for publication in this renowned publishing company and accompanied the preparation of this edition with many very helpful suggestions. The detailed work of the production of this book was in the hands of Lesley Poliner, Anne Fossella and David Kramer to all of whom we are very much obliged, in particular to Lesley Poliner for her competent and kind cooperation and patience with the authors.

Rudolf Fritsch and Gerda Fritsch
München, Germany
May 11, 1998

Ways to Read This Book

First and foremost everyone should read this book out of personal interest! Chapter 1 contains the historical development in so far as can be presented without a wide range of conceptual explanations of a mathematical nature. In addition, it presents particular biographical information about the lead players in this mathematical drama. The succeeding chapters, too, touch time and again on the historical backdrop but are more oriented towards the germinating ideas and their formulations and less to the personalities themselves. Chapter 2 deals with the topological foundations. The ideas in this chapter are particularly important for those who ultimately seek a deeper understanding of the material. One should attempt, at least, to master the statements of the theorems and their meanings. The proofs of the theorems and the often highly technical lemmas one can, on the first reading, gloss over or simply omit.

Those who want very quickly to get to the heart of the Four-Color Theorem and the technical difficulties associated with it can begin reading at Chapter 3. There one finds a precise formulation of the topological version of the theorem and a few of its related problems. As to understanding the proofs, it suffices to look up the relevant concepts that have been defined previously in Chapter 2.

A reader who, on the other hand, is familiar with the Euler characteristic and has a basic understanding of graph theory may begin reading this book with Section 6 of Chapter 4. This section contains the entire foundational material and all of the important mathematical formulae required for the proof of the Four-Color Theorem. Sections 4.7 and 4.8 can be skipped. A thorough reading of Chapters 5, 6, and 7 is advisable for whoever wants to gain an impression (and an appreciation) of the technicalities of the proof of the theorem itself.

Contents

1

CHAPTER

History

Only a few mathematicians will have heard of Francis Guthrie, an able lawyer and recognized amateur botanist in whose honor the South African flora *Guthriea capensis* and *Erica Guthriei* were named. However, every mathematician knows of the Four-Color Theorem as a very interesting and, many think, still unsolved mathematical problem with which a long line of the best minds in this discipline have grappled.

What does the one have to do with the other?

Normally, the statements of mathematical conjectures have an "inventor"—that is, someone who, at some time or other, expressed this particular query of interest as a concrete mathematical statement for the very first time. It is rare for mathematical conjectures to be formulated merely because definitive answers to them would lead to the solution of some very practical problem. Moreover, there are naturally only a few problems of interest to mathematicians that are capable of being formulated in such a way as to be understood by virtually anyone.

One notable exception to this is the Four-Color Problem. Its initiator was Francis Guthrie, a man who lived during the middle of the nineteenth century. Guthrie was not only a lawyer and a botanist. He was, in fact, primarily a mathematician.

1

FIGURE 1.1   Francis Guthrie

One day, as he was coloring a map of England, he thought he had discovered that one could always color an arbitrary map with four colors, given the requirement that countries with common borders be colored differently.

Guthrie, Francis

* *London 22.1.1831,* † *Claremont, South Africa, 19.10.1899.*
Francis Guthrie studied mathematics at University College,[1] in London,

[1]The secular University College London was established on October 1, 1828, and ran until 1836 under the name "London University." It was from the outset open to students of all denominations, in contrast to the older universities in Oxford and Cambridge. Because of that, three years later, on October 8, 1831, King's College London opened, leaning strictly towards the Anglican Church. In 1836 the "University of London" was set up as an examining and degree-granting body having authority over both Colleges. In 1880 all academic degrees, distinctions,

where he obtained a Bachelor of Arts (B.A.) degree in 1850. Subsequent to that, he decided to study law, and he finished his Bachelor of Laws (LL.B.) degree in 1852. Guthrie passed the final examinations for both disciplines with first-class honors and was made a Fellow[2] of University College in 1856, a title he held until his death. He was first admitted to the "Middle Temple" Inn of Court and later, in 1857, was called to the Bar.[3] For a few years after that he was active in consulting and chancery practice.

In April 1861, Guthrie opted for a professorship in mathematics at the newly established Graaff-Reinet College in the Cape Colony of South Africa. As an academic, he was reputed to be a very hard worker but was characterized as being warm-hearted, full of humor, patient, and unassuming. Personal concern for his students who were truly interested in mathematics was, for him, more important than his lectures. His balanced perspective contributed to a positive, productive atmosphere around the College.

He also gave highly successful public lectures in botany, which were inspired by the famous Mr. John Lindley, a man with whom he had been acquainted during his student days in London. Otherwise, he busied himself with the new railroad that was to link Port Elizabeth to the Indian Ocean. It was to be located in the region behind the mountain range situated near Graaff-Reinet. He used to climb the mountains in that area with friends in order to discover the best possibility for the construction of a pass through the mountains for the railway lines. The railroad tracks

---

and prizes were made accessible to women. By 1898 the University of London was not only an examining body but had also become a teaching institution.

[2]In the English university system of the last century, fellows were the highest qualified members of a scholarly organization, for example a college. There existed several kinds of fellowships. They were a kind of sinecure whereby the bearers of this designation were under varying obligations (some carried no duties at all) to their college.

[3]Barrister was the professional designation for an English lawyer who was permitted to plead cases as an attorney-at-law in the higher Courts of Justice. After at least a three-year affiliation with one of the four London Inns of Court, a barrister would be called to the Bar by the governing board, the so-called Bench. The London Inns of Court were the actual training centers for English lawyers up until the middle of the nineteenth century.

themselves were also built in accordance with the recommendation of Guthrie and his friends.

In 1875 he left his position in Graaff-Reinet and moved to Cape Town. For a short time in 1876, he and his friend, Harry Bolus, a businessman and botanist, traveled together to England where, at the famous Kew Gardens, both pursued botanical studies on the flora of the Cape region.

A year later, in 1877, Guthrie became one of the founders of the South African Philosophical Society, later called the Royal Society of South Africa. It was to this society that he, on November 28 of that same year, delivered a lecture entitled "The Heat of the Sun in South Africa" on a topic that is current even today. In the introduction, he stated that it must be possible to convert the energy of the sun into mechanical power. Moreover, he optimistically expressed the opinion that the technical difficulties still standing in the way of economic exploitation of the almost uninterrupted hours of sunshine in South Africa would, in a few years, be overcome. The subsequent actual execution of his ideas is today still only of historical interest.

Guthrie also seemed to take interest in the development of the first aircraft. He was referred to as the inventor of a flying machine. However, nothing specific regarding that could be found in the literature.

In November 1878, Guthrie accepted a professorship in mathematics at South African College, in Cape Town. He held this position until his retirement, due to illness, in January 1899.

He was a member of the Cape Meteorological Commission and, in 1878/79, general secretary of the South African Philosophical Society.

His mathematical papers are not outstanding, from the points of view of both content and quantity. Those with which we are familiar deal mainly with questions from elementary algebra and with technical problems from physics.

Other than through his connection to the Four-Color Theorem, he earned an enduring name for himself with his botanical research into Cape flora and the categorization of heathland flowers. Bolus had encouraged him to devote even more of his time to botany and had honored him by using Guthrie's name on two separate occasions for botanical nomenclature:

- Guthriea capensis *was one of the newly discovered (by Bolus in 1873) types of plants out of the Achariaceae family in the Gnadouw Snow Mountain region, and*

● Erica Guthriei *was a species out of the Ericaceae family, newly discovered at Piquet Mountain by Guthrie himself.*

In 1967 there appeared a monograph entitled "Ericas in Southern Africa," in which the article "Pioneers in the Study of Ericas in Southern Africa" was dedicated to the memory of Francis Guthrie and Harry Bolus [BAKER and OLIVER 1967].

Guthrie died nine months following his retirement and was buried in the cemetery of the St. Thomas Anglican Church in Rondebosch. His grave is still maintained. His herbarium was donated by his widow to South African College. In 1983 it was integrated into the larger Bolus Herbarium (also located at the University of Cape Town). The two cabinets housing the Guthrie Herbarium have been preserved there in a separate enclave under its original name.[4]

Francis Guthrie had a younger brother, Frederick Guthrie, who also studied at University College. It was Frederick who submitted Francis's observation ("*With my brother's permission . . .*") as a mathematical conjecture to his professor Augustus de Morgan on October 23, 1852.

De Morgan was fascinated by the Four-Color Problem and, on the very same day, wrote a letter to Sir William Rowan Hamilton in which he described the situation and explained the problem. This letter has been preserved and today is kept in Trinity College, Dublin [DE MORGAN 1852]. It is the very first written correspondence regarding the Four-Color Problem. On the following page is a facsimile of the relevant passages and, across from it, a transcript of them.

In contrast to de Morgan, Hamilton did not think the problem at all interesting. Four days later (as is indicated by his correspondence), he responded by stating that he would not soon be tackling de Morgan's "quaternion of colours" problem.

In the time that followed, it was through de Morgan that the Four-Color Problem, as a mathematical conjecture, was much more extensively made known—so much so that he is considered by many to be the actual originator of the problem. The first time that the mathematical world learned of the true originator of the Four-Color

---

[4]This information was kindly provided by Patricia Lorber, the present curator of the Bolus Herbarium.

Problem was in the article "Note on the Colouring of Maps" [GUTHRIE 1880], published by Frederick Guthrie in 1880 in the *Proceedings of the Royal Society of Edinburgh*. In this memorandum, Frederick Guthrie himself maintained that de Morgan had always acknowledged the one who had first informed him of this problem.

Guthrie, Frederick

∗ London 15.10.1833, † London 21.10.1886.
After his studies in London, Francis Guthrie's younger brother, Frederick, went to Heidelberg, where he attended lectures given by Robert Bunsen. Subsequent to that, in 1854, he was granted a Ph.D. from the University of Marburg. One year later, in 1855, after his return to England, he acquired a bachelor of arts degree from University College, in London. His academic career included positions in Manchester and in Edinburgh. In 1861 he became professor of chemistry at the Royal College of Mauritius, a position he held for eight years. He was also a professor of physics, first at the Royal School of Mines, in 1869, and later at the School of Sciences, in London, in 1881. He was founder of the Society of Physicists of London and became its president in 1884. From 1860 onwards, he was a fellow of the Royal Society of Edinburgh and, in 1873, became a fellow of the Royal Society of London. He also published several poems under the pseudonym Frederick Cerny.

de Morgan, Augustus

∗ Madura, India 27.6.1806, † London 18.3.1871.
De Morgan studied mathematics in Cambridge and, in 1827, passed his bachelor's exam as "fourth wrangler."[5] In order to ensure a means of liveli-

---

[5]In Cambridge, the term "wrangler" applied to those who passed the tripos mathematics exam, thereby belonging to the upper half of the candidates. The best one was called senior wrangler; then came the second wrangler, the third wrangler, and so on. (The word wrangler is also suggestive of the, up to 1839, customary debating exercises in Cambridge.)

My dear Hamilton

A student of mine asked me to day to give him a reason for a fact which I did not know was a fact — and do not yet. He says that if a figure be any how divided and the compartments differently coloured so that figures with any portion of common boundary line are differently coloured — four colours may be wanted but not more — The following is his case in which four are wanted

A B C & D are names of colours

Query cannot a necessity for five or more be invented. As far as I see at this moment, if four ultimate compartments have each boundary line in common with one of the others, three of them inclose the fourth, and prevent any fifth from connexion with it. If this be true, four colours will colour any possible map without any necessity for colour meeting colour except at a point.

Now it does seem that drawing three compartments with common boundary A B C two and two — you cannot make a fourth take boundary from all, except by inclosing one — But it is tricky work and I am not sure of all convolutions — What do you say? And has it, if true been noticed? My pupil says he guessed it in colouring a map of England

B is inclosed

The more I think of it the more evident it seems. If you retort with some very simple case which makes me out a stupid animal, I think I must do as the Sphynx did. If this rule be true the following proposition of logic follows

If A B C D be four names of which any two might be confounded by breaking down some wall of definition, then some one of the names must be a species of some name which includes nothing external to the other three

Yours truly
A De Morgan
Oct 23/52.

Extract from de Morgan's original letter to Hamilton,
printed with kind permission from
The Board of Trinity College, Dublin.

My dear Hamilton
:

A student of mine asked me to day to give him a reason for a fact which I did not know was a fact, and do not yet. He says, that if a figure be any how divided and the compartments differently coloured so that figures with any portion of common boundary <u>line</u> are differently coloured – four colours may be wanted but not more. The following is his care in which four <u>are</u> wanted.

A B C D are names of colours

Query cannot a necessity for five or more be invented. As far as I see at this moment, if four <u>ultimate</u> compartments have each boundary line in common with one of the others, three of them inclose the fourth, and prevent any fifth from connexion with it. If this be true, four colours will colour any possible map without any necessity for colour meeting colour except at a point.

Now, it does seem that drawing three compartments with common boundary A B C two and two – you cannot make a fourth take boundary from all, except inclosing one – But it is tricky work and I am not sure of all convolutions – What do you say? And has it, if true been noticed? My pupil says he guessed it in colouring a map of England. The more I think of it, the more evident it seems. If you retort with some very simple case which makes me out a stupid animal, I think I must do as the Sphynx[b] did. If this rule be true the following proposition of logic follows:–

If A B C D be four names of which any two might be confounded by breaking down some wall of definition, then some one of the names must be a species of some name which includes nothing external to the other three.

Yours truly

A De Morgan

Oct 23/52

---

[b] The Sphinx of ancient Greek Mythology who plunged to her own death when Oedipus solved a difficult riddle she had given him.

FIGURE 1.2   Augustus de Morgan

hood for himself, he subsequently decided to study law and was admitted to Lincoln's Inn.[6] In 1828, he became the first professor of mathematics at the newly established University College, in London. Three years later (in 1831), however, he relinquished his professorship there because he realized, after the dismissal of a colleague of his, that academic freedom would no longer be safeguarded by the college administration. Nonetheless, in 1836, he returned to the college. He retired in 1866 when the University College administration refused to hire a highly qualified applicant simply because of his Unitarian beliefs.

De Morgan was one of the founders of the London Mathematical Society and was its first president. He was purported to be a first-rate teacher whose lectures attested to his scientific originality, his humor, and his didactic skills. His general liberal attitude led him, in later years, to regard women's studies in an increasingly more favorable light.

De Morgan's scholarly achievements are found in monographs, countless articles in various journals, and some 650 contributions to an encyclopedia. His mathematical work centers mainly on analysis and

---

[6]Another of the London Inns of Court (see footnote 3).

logic. In particular, the "De Morgan Laws" in logic are named after him. He and George Boole, the founder of formal logic, maintained an extensive correspondence, which has been preserved for posterity.[7]

Hamilton, Sir William Rowan

* Dublin 4.8.1805, † Dunsink 2.9.1865.

William Hamilton was regarded as a child prodigy with an extraordinary aptitude for languages. By the time he was 14, he had mastered more than a dozen languages (including several Oriental and Asiatic ones). He did not attend school until he enrolled in Trinity College, Dublin, at 18. There he performed brilliantly, not only in the classics but also in mathematics.

In 1827, when he was just 22 years old and before he had graduated, he was made the Andrews Professor of Astronomy at Dublin University and the Royal Astronomer of Ireland. In his private life, he was not as fortunate, having had a somewhat disruptive home life. It was this situation that in his later years resulted in his turning increasingly to alcohol for solace.

The discovery of the quaternions was his most sensational mathematical achievement. However, contrary to what was thought at the end of the last century, they indeed do not exclusively form the foundations of mathematics. They do, however, give an essential insight into the structure of number fields.

Hamilton was a fellow of numerous European scientific societies. Shortly before his death, the American Academy of Science nominated him as their first non-American member. He is also regarded as the greatest scientist that Ireland has ever produced.

The next traditional sources concerning the Four-Color Problem date from the following two years, 1853 and 1854. They are contained in two letters: The first, de Morgan wrote to his friend and former teacher William Whewell on December 9, 1853; the second was a letter from de Morgan to Whewell's would-be brother-in-law Robert Leslie Ellis, dated June 24, 1854. Both are preserved in Trinity

---

[7]A somewhat comprehensive biography of Augustus de Morgan can be found in [FRITSCH 1991].

College, in Cambridge [DE MORGAN 1853, 1854]. In these two letters, de Morgan discusses the Four-Color Conjecture. The question that occupied him the most was whether one could prove that in the case of four countries that pairwise have a common border, one of them would be enclosed by the other three. Mistakenly he took that as the core problem.

On April 14, 1860, in the journal *Athenæum*, there appeared a review of William Whewell's book *The Philosophy of Discovery* [DE MORGAN 1860]. In it was found an anonymous (which was customary for Athenæum) reference to the Four-Color Problem described precisely in the way outlined above. The proof that de Morgan had written the review comes from a letter[8] dated March 3, 1860, in which de Morgan expressed his thanks to the author Whewell for sending him a copy of his book. In it he mentioned that he had also received a copy from Athenæum. This he had apparently sent back unopened. Future generations of editors would no doubt surely have wondered how reviews could have been written if the referees had never once read the books.

On the basis of this book review (*"Now, it must have been always known to map-colourers that* four *different colours are enough"*), probably another unfounded assertion emerged, namely, the belief that cartographers at that time had acknowledged the statement of the Four-Color Problem from their professional practice. However, there exists no scientific documentation for that.

Modern cartographers, such as Manfred Buchroitner, in Dresden, and Hans Günther Gierloff-Emden, in Munich, have even mentioned the Four-Color Problem in their lectures—however, not as a scientific cartographical formulation but more as a somewhat curious historical footnote. Kenneth O. May, who in the 1960s was interested in the history of the Four-Color Problem, examined the huge atlas collection in the Library of Congress, in Washington. He found no evidence of any kind of a minimal coloring in the production of maps. The idea that some mapmakers had at one time or

---

[8]It is printed in the book *Memoirs of Augustus de Morgan*, which Sofia Elizabeth de Morgan, the wife of Augustus de Morgan, wrote and which appeared in 1872 [DE MORGAN S.E. 1872].

other labored over the Four-Color Problem is misleading because in cartography, a restriction to only four colors has never been necessary.

Whewell, William

∗ *Lancaster 24.5.1794,* † *Cambridge 5.3.1866.*
William Whewell began his studies in 1812 at Trinity College, in Cambridge, a college he remained associated with for his entire lifetime. In 1828 he became professor of mineralogy, a position he relinquished in 1833. From 1838 to 1855, he was professor of ethics; he was college master in the years 1841 to 1866, and in the academic years 1842/43 and 1855/56 he was vice-chancellor of the university.

His publications encompass (which was typical for many scholars in the nineteenth century) a broad spectrum of subjects: mechanics and tidal phenomena, an in-depth analysis of the works of Kant, translation of German literature, and the history of (including questions concerning) education. His first works dealt with mathematical issues and resulted in a radical reform of the teaching of mathematics at Cambridge University.

Whewell's handbook on statistics and dynamics went through several editions. His best-known works are the History of the Inductive Sciences, *published in 1837, and the two-volume* Philosophy of the Inductive Sciences *of 1840.*

Ellis, Robert Leslie

∗ *Bath 25.8.1817,* † *Anstey Hall, Trumpington 12.5.1859.*
Robert Ellis went to Trinity College, Cambridge, in October of 1836 and was, in his last semester, a student of William Hopkins. In January 1840 he passed his bachelor examinations as "senior wrangler." In October of the same year he was nominated as a fellow of Trinity College. Ellis was primarily interested in civil rights. However, due to an inheritance, he did not need to work as a lawyer. After the death of Duncan Farquharson Gregory, the founder of the Cambridge Mathematical Journal, *Ellis became editor of sections of its third and fourth volumes. This journal was*

the forerunner of the present Quarterly Journal of Pure and Applied Mathematics.

Ellis's name has also been linked with the complete works of Francis Bacon.

After 1860, for a period of about twenty years, the interest of mathematicians in the Four-Color Problem appeared to wane. The problem was not discussed anywhere in the mathematical literature of the time. It was, nonetheless, obviously not forgotten, as is illustrated by an inquiry of one Arthur Cayley on July 13, 1878, in the mathematical section of the Royal Society [LONDON MATHEMATICAL SOCIETY 1878]. Cayley wanted to know whether anyone had as yet submitted a solution to the Four-Color Conjecture. With that, the problem was again put back into the spotlight for the mathematical community. Cayley himself published a short, basic analysis of the problem in the *Proceedings of the Royal Geographical Society* [CAYLEY 1879].

FIGURE 1.3 Arthur Cayley

Cayley, Arthur

∗ Richmond, Surrey (today part of the City of London) 6.8.1821,
  † Cambridge 26.1.1895.
*Cayley began his studies at Trinity College, Cambridge, in 1838. He was
an outstanding and talented student, and he emerged from the mathematical tripos[9] examinations as "senior wrangler" and "Smith's Prizeman."[10]
Because no position in mathematics was forthcoming at the time, he was
admitted to Lincoln's Inn in 1846 and was called to the Bar in 1849.
Cayley was a brilliant lawyer who did well in his profession. However,
he limited the scope of his law practice somewhat so that he had time to
pursue his mathematical interests. This was reflected in the publication
of some three hundred mathematical papers that appeared during his
fourteen years as a lawyer.*

*In 1863 Cayley was granted the newly established Sadlerian chair
at the University of Cambridge. He taught there until his retirement in
1892, a period broken only by a lecture tour to Johns Hopkins University,
in Baltimore, in the winter semester of 1881/82.*

*Cayley was granted innumerable distinctions and academic honors
throughout the entire world. From both university and scientific circles,
his advice in matters of law was often sought. He was described as a
modest and extremely reserved man.[11]*

*Cayley, together with Sylvester, is thought to be an originator of the
theory of invariants. At the beginning of this century, this theory was
considered outdated, but in recent times it has experienced an enormous
reawakening of interest. Its fundamental tenets were formulated while the
two lawyers took walks about the law courts during the recesses of the
judicial hearings and used the free time to discuss mathematics. Apart*

---

[9]See footnote 4.

[10]The prestigious Smith's prizes, to the amount of £25, were awarded twice a year
at Cambridge University for top performance in mathematics and the natural
sciences.

[11]A few of his interests seem from today's standpoint very modern. He loved traveling, in particular, for the purpose of acquainting himself with architectural styles
and works of famous artists. Already as a young man he had traveled extensively
on the continent. He was an enthusiastic mountain climber and quickly became
a member of the Alpine Club. He also dedicated himself to better educational
standards for women.

from that, Cayley developed matrix calculus and provided outstanding contributions to algebraic geometry, a subject many of whose concepts are named after him.

In a short paper that he had drafted when he was a young man [CAYLEY 1848] and that had long been in the making, he introduced chain complexes and defined homology and Reidemeister torsion. He discussed these notions very explicitly in terms of sequences of multipliable matrices and also offered his ideas about their rank and corank. An abstract version of these topics is now a part of the foundations of algebraic topology, a subject that was developed only in the first half of the twentieth century.

Almost a year to the day after Cayley's article regarding the Four-Color Problem appeared in the journal *Nature*, "*the solution of a problem which recently achieved some renown*" (Cayley) would be found by a man named Alfred Bray Kempe, and the complete proof of this problem would soon appear in the *American Journal of Mathematics*, a publication that had been founded in 1878.

James Joseph Sylvester was the founder and editor-in-chief of the journal. Like his friend Cayley, he was both a lawyer and a mathematician and, from 1876 on, a professor at the Johns Hopkins University, in Baltimore. It is no doubt due to the close friendship between Cayley and Sylvester that Kempe—"*on the request of the editor-in-chief*" of the *American Journal of Mathematics Pure and Applied*—submitted his outstanding and significant paper to a comparatively insignificant (at the time) American publication.

With Kempe, yet another lawyer had stepped into the limelight of the Four-Color Problem.

The associate editor of the American journal, William Edward Story, subsequently added a few addenda to the reprint of Kempe's article [KEMPE 1879a]. These addenda treated special cases not originally included by the author [STORY 1879]. It was in this form that Story presented the entire paper at a meeting of the Scientific Association at Johns Hopkins University on November 5, 1879.

The illustrious American philosopher, logician, and mathematician Charles (Santiago) Sanders Peirce, who happened to be teaching in Baltimore in 1879, added a few observations of his own, and in the next session of the association, on December 3, he ex-

FIGURE 1.4   Sir Alfred Bray Kempe

plained how, in his opinion, Kempe's proof could be improved using rules of logic [SCIENTIFIC ASSOCIATION 1880].[12]

Thus, in 1879, the Four-Color Problem was considered to have been solved.

Kempe, Sir Alfred Bray

∗ *Kensington (today also a part of the City of London) 6.7.1849,*

---

[12]In an unpublished manuscript, which is preserved in the Houghton Library of Harvard University, Peirce alleged that he had, in the 1860s, lectured to a mathematical society of the university on one attempt at proving the Four-Color Theorem. This had probably been inspired by de Morgan's book review in *Athenæum*. Peirce remained interested in the Four-Color Theorem all his life [EISELE 1976]. In the history of this problem, he is the only mathematician who has also been involved with cartography. The results of his work in that field are still implemented today in the depiction of international flight paths.

† *London 21.4.1922.*

Alfred Kempe studied at Trinity College, in Cambridge, where he also attended Cayley's lectures. He graduated with a B.A. degree as "22nd wrangler." He was then admitted to the Inner Temple[13] and, in 1873, completed his training as a barrister. Following that, he joined the "Western Circuit." The circuits, of which there were eight in total throughout England and Wales, were organizations for the administration of justice. They dated back to the 12th century. Three times per year, two judges of the High Court of Justice with an entourage of lawyers and a few prestigious gentlemen of the respective shires set out on a circular tour of various parts of the country to act as a jury court.

Kempe then offered his legal expertise as a service to the Anglican Church. Among other things, he was Secretary of the Royal Commission for Ecclesiastical Jurisdiction (1881 to 1883). In addition, he was chancellor of the following dioceses: Southwell and Newcastle (both in 1887), St. Albans (1891), and London (1912). However, the lawyer Kempe also made himself a name in mathematics and, in 1881, was made a fellow of the Royal Society for his work on linkages. Other mathematical works of his that at that time were significant were involved primarily with knot theory and lattice theory, and also with mathematical logic.

Besides his involvement with the Church, Kempe for many years served in an honorary capacity in scientific associations. He was treasurer of the London Mathematical Society and, in 1893/94, its president. From 1899 to 1919 he belonged to the Board of Governors of the Royal Society. From time to time, he was also on the Treasury Board and was elected its vice-president. Moreover, he was treasurer of the National Physical Laboratory and managing director of the Royal Institution of Great Britain, an association founded for the cultivation and broadening of scientific knowledge. His expertise in legal matters found further recognition in 1909 when he was elected to the Bench of the Inner Temple.[14]

Kempe was knighted in 1912. In a dictionary of the time one finds the following comment: that he relaxed by indulging in mathematics and music.

---

[13] The third Inn of Court. See footnote 3.
[14] See footnote 3.

Sylvester, James Joseph

∗ London 3.9.1814, † London 15.3.1897.

James Sylvester finished his studies in Cambridge in 1837 as "second wrangler" but did not obtain an academic degree because due to his Jewish faith he could not subscribe to the 39 article of the Constitution of the Anglican Church.[15]

Sylvester's academic career was out of the ordinary and many-faceted. For differing lengths of time he was a professor in London, Virginia, Woolwich, Baltimore, and Oxford. Between the professorships, there were several involuntary interludes in which Sylvester worked as an actuary and during which he began his law studies at the Inner Temple. This culminated in his being called to the Bar in 1850.

In 1878, while he was teaching at the Johns Hopkins University, in Baltimore, Sylvester, together with Story as associate editor-in-charge, founded the American Journal of Mathematics.

Despite the close friendship with Cayley,[16] *Sylvester was of a completely contrary disposition: uncontrollably impetuous, often erratic in his mathematical thinking, concentrating only on what occupied him at the moment, and completely uninterested in the mathematical thoughts of others.*

He was considered to be one of the greatest mathematicians of the 19th century. Charles Sanders Peirce said of him: "He was perhaps the mind most exuberant in ideas of pure mathematics of any since Gauss."

Story, William Edward

∗ Boston 29.4.1850, † Worcester, Massachusetts, 10.4.1930.

William Story studied mathematics and physics at Harvard University,

---

[15]Sylvester nevertheless acquired his B.A. and his M.A. in Dublin in 1841. After the Test Act of 1872, he also received both previously ungranted degrees *honoris causa* from Cambridge.

[16]Both have coined many new mathematical words that have Greek or Latin derivatives. The word "graph" is one such example. Sylvester used this word for the first time in 1878 [SYLVESTER 1878, page 284]. Due to his countless neologisms, Sylvester himself described himself as a "Mathematical Adam."

taking his final examinations there in 1871. Following that, he furthered his studies in Berlin, where among other things, he attended lectures by Karl Weierstrass and Ernst Kummer. He obtained his doctorate under Carl Gottfried Neumann in Leipzig in 1875. A year later, when the Johns Hopkins University was opened in Baltimore, Story obtained an assistant professorship there and later became associate professor.

Two years after that, when Sylvester founded the American Journal of Mathematics, *Story became its associate editor-in-charge.*

From 1889 until his retirement in 1921, he was professor at Clark University, in Worcester. He was a member of the American Academy of Arts and Sciences and was appointed to the National Academy of Science.

> *A bit of family history: His great grandfather Elisha Story was one of the citizens who, on December 16, 1773, at the so-called Boston Tea Party, dressed up as Indians and destroyed a boatload of tea from the East India Trading Company in the Boston harbor.*

Peirce, Charles (Santiago) Sanders

∗ Cambridge, Massachusetts, 10.9.1839,
  † Milford, Pennsylvania, 19.4.1914.
Peirce was the son of the well-known mathematician and lecturer Benjamin Peirce. In 1861, after his studies in mathematics at Harvard University, Peirce became a coastal surveyor for the United States, and he worked in that profession for the next thirty years.

Peirce did not manage to establish an academic career for himself. In total, he taught for only eight years at Johns Hopkins University, at Harvard University, and at the Lowell Institute, in Boston. His lectures were very difficult to understand and were addressed only to the very gifted students whom he often confronted in a patronizing way that exhibited a lack of self-control. For the last twenty years of his life, he lived in complete seclusion, in part in such poor financial circumstances that he had to rely on the financial assistance of kindly disposed friends.

Peirce has been considered to be the founder of the philosophical school of pragmatism to which some of the foremost American philosophers of the day ascribed. His greatest contribution was, nevertheless, in the field of logic, where he developed, among other things, proportional calculus. His numerous papers, of which many exist only as manuscripts, were not accessible to a broad readership until the beginning of the twentieth century, when logic was first established as an academic subject at the university level.

In 1880 the physicist Peter Guthrie Tait published a new (in his opinion) proof of the Four-Color Problem in the *Proceedings of the Royal Society of Edinburgh* [TAIT 1880]. Tait had, nonetheless, found only a few interesting reformulations of the problem, and the lawyer Kempe was still considered the one who had proven the Four-Color Theorem.

Tait, Guthrie Peter

* Dalkeith, Scotland, 28.4.1831, † Edinburgh 4.7.1901.
Guthrie Tait began his studies at the University of Edinburgh. However, a year later, in 1848, he went to Peterhouse College, in Cambridge, where he graduated in 1852 as "senior wrangler" and as "first Smith's Prizeman."[17] *In 1854 Tait, already a fellow of his college, left Cambridge and became a mathematics professor at Queen's College, in Belfast. Successful collaborative work in the area of ozone research and in thermoelectricity linked him with his colleague Thomas Andrews, an outstanding Irish chemist. In 1860 Tait obtained a professorship in natural philosophy in Edinburgh, a position he held until shortly before his death.*

It was in Edinburgh, in coauthorship with Sir William Thomson (later Lord Kelvin), that his most famous work was written. This was his Treatise on Natural Philosophy, *which came to be known as "T and T'." It was published in Oxford in 1867. It became a classical text for mathematical physics, in which the principle of the conservation of energy was traced back to Isaac Newton.*

Tait's scientific achievements lay mainly in the area of experimental physics, with emphasis in thermodynamics, thermoelectricity, and kinetic gas theory. He had, however, in addition, a place of significance in

---

[17]See footnote 8.

the development of mathematics. On the one hand, he was champion for Hamiltonian quaternion calculus, out of which he wanted to establish a kind of vector analysis. On the other hand, he developed interesting conjectures in knot theory. According to an announcement made in 1991, the last of his conjectures has just recently been proven [MENASCO and THISTLETHWAITE 1991].

The solution of the Four-Color Problem by an outsider probably led many to assume that it could not have been a matter of anything very profound, despite its initial celebrated rise into the limelight. Perhaps Felix Klein was also seduced into that way of thinking—for he related a fairy tale about it that has been repeated time and again, even up to the present day, indicating that the Four-Color Theorem, in a moderately altered form, had already been discussed around 1840 in August Ferdinand Möbius's lectures.

The backdrop for this bit of history is the following. Felix Klein had been present at the January 12, 1885 meeting of the mathematical-physical class of the Leipzig Scientific Society. The geometer Richard Baltzer reported in that meeting about a remark in one of Möbius' lectures and a fitting discovery in the papers of Möbius estate:

> In 1840, in a lecture in which geometry assignments were being discussed, Möbius related this to his listeners:
>
> *"Once upon a time, a king in India had a large kingdom and five sons. His last will and testament decreed that, after his death, the sons should partition the kingdom in such a way that each one's region would have a common boundary (not merely a single point) with each of the other's regions. How was one to divide the kingdom?"*
>
> As we would in the next lecture acknowledge, we had strived in vain to find such a partition. Möbius chuckled and remarked that he was sorry that we had gone to so much effort because the required partition was, in fact, impossible [BALTZER 1885].

By examining the written works in the estate, Baltzer discovered that Möbius had obtained his wisdom from his friend, the philologist Benjamin Gotthold Weiske. Baltzer then added an observation to the published form of the lecture [BALTZER 1885]:

My colleague Klein had the goodness to draw my attention to the work of Mr. Kempe in London which alludes to the aforementioned topic. According to the above lemma about the partitioning of the kingdom, the fact under discussion, namely the Four-Color Theorem, is immediately clear. How delighted Möbius would have been if he had known the remarkable economic significance of the result communicated to him by his friend Weiske.

It was at this point that Baltzer had made an error. Weiske's theorem, indeed, made possible a simple proof of the *Five*-Color Theorem, but it contributed absolutely nothing to the proof of the Four-Color Theorem.

Nevertheless, the Möbius–Weiske tale spread far and wide. In Germany, Friedrich Dingeldey [DINGELDEY 1890] was, in part, responsible for that. A widely read memorandum by Isabel Maddison also carried the fairy tale to North America [MADDISON 1897]. It was not until 1959 that H.S.M. Coxeter brought the error to light [COXETER 1959].

After the publication of Kempe's proof, the furor surrounding the Four-Color Problem, apart from Tait's attempt at proving it, faded away into silence.

This silence was suddenly shattered in 1890 when Percy John Heawood showed that Kempe's proof contained a fallacy!

Heawood explained, to his regret, that he had no proof for the Four-Color Theorem and that his article was, unfortunately, not constructive for he had only found a mistake in the up to that time accepted proof [HEAWOOD 1890].

Heawood, Percy John

* *Newport, Shropshire, 8.9.1861,* † *Durham 24.1.1955.*
Percy Heawood studied at Exeter College, in Oxford, where he received his bachelor of arts degree in 1883 and his master of arts in 1887. In addition, he studied, with a good deal of success, classical languages, which he enjoyed reading from the original texts throughout his life.

In 1887 he became lecturer of mathematics at Durham College, which later became the University of Durham. There, in 1911, he obtained a

FIGURE 1.5   Percy John Heawood

professorship in mathematics and retired only in 1939 at the age of 78. From 1926 to 1928 he was vice-chancellor of the university. Before and after that he held various other university administrative positions.

Numerous articles in the Mathematical Gazette, *a journal dedicated to the teaching of mathematics, testify to the fact that Heawood was very interested in promoting a good relationship between secondary and post-secondary education. He was also the first chairman of the University of Durham Schools Examination Board. The so-called Examination Boards were set up in the second half of the previous century in a number of English universities to aid students in obtaining their diplomas.*

He was a devout member of the Anglican Church and throughout his life held numerous official positions in the Church as a lay person.

He received two important honors, not for mathematical merit, but for his untiring and unparalleled efforts towards the reconstruction of Durham Castle. The castle, a historical structure of utmost value, had begun being constructed in the 10th century, and by the 1860s it had started to slide over the cliffs. The university did its best to save it but finally gave up because of the horrendous cost required to maintain it. Heawood campaigned, as secretary of the Restoration Fund, practically alone year after year until the castle could again be restored. For that he received an honorary doctorate in 1931 from the University of Durham, and in 1939 he was conferred the title of Order of the British Empire.

Heawood was an unusual persona. Already during his lifetime there were many anecdotes about him. He remained in Durham until his death at almost 94 years of age.

It is not known how Heawood became aware of the Four-Color Problem. However, his life's work in mathematics was centered on it. After his first paper, in 1890, he published seven more papers on the same subject, the last in 1949 at the age of 88. His biographer, Gabor Dirac, exaggerated when he claimed: "... his discoveries are more substantial than all later ones by all others put together" because, with hindsight, one can make the observation that the final proof of the Four-Color Theorem by Appel, Haken, and Koch used relatively few of Heawood's ideas. Nonetheless, Heawood remains a central figure in the history of the Four-Color Problem.

Heawood's article in which he refuted Kempe's proof of the theorem had a positive outcome—namely, the proof of the above-mentioned Five-Color Theorem.

In addition, Heawood considered the question, already put forward by Kempe, of the minimum number of colors required for an admissible coloring of maps of the sphere and other closed surfaces, such as the torus and the surface of genus 3.

Heawood gave an upper bound for this. The problem regarding the exactness of this upper bound is called the *Heawood map-coloring problem* and, surprisingly, was easier to solve than the Four-Color Problem. Gerhard Ringel and J.W.T. Youngs supplied a significant contribution to the solution in a fortuitous and almost breathtaking race for time with Jean Mayer. Mayer solved the difficult case for $n = 30$ between February 25 and 27, 1968. Ringel and Youngs, on the other hand, disposed of the cases $n = 35, 47, 59$ on March 1 and the case $n = 30$ on March 4, 1968 [RINGEL and YOUNGS 1968; MAYER 1969].

In the limited scope of this book we are not able to discuss this matter at any greater length. Instead, we refer you to the pertinent literature [RINGEL 1974, AIGNER 1984].

In the years following Heawood's discovery, the Four-Color Theorem would be discussed at great length but to no great success. In retrospect, William Thomas Tutte identified two different approaches: one "qualitative," the other "quantitative" [TUTTE 1974].

The qualitative approach further pursued the ideas behind Kempe's method, the so-called "Kempe chain games." The quantitative approach continued along the lines of Heawood's ideas.

In his second work, Heawood tackled the Four-Color Problem with methods from elementary number theory [HEAWOOD 1897]. Oswald Veblen took up these notions and delivered a presentation of the problem in the form of linear equations over a finite space to the American Mathematical Society on April 27, 1912 [VEBLEN 1912].

Veblen had a somewhat younger colleague named George David Birkhoff. In his seminars in Princeton, Birkhoff and his student and collaborator Philip Franklin entrusted the problem to all the topologists there. Birkhoff's first work in this regard was directly related to what Veblen had done [BIRKHOFF 1912] and thereby ascribed to the quantitative method. In his second paper, he went back to Kempe's qualitative approach and achieved a decisive step forward with the notion of *reducible rings*—a notion to which earlier authors, for example Paul Wernicke [WERNICKE 1904], had only alluded.

Birkhoff, George David

* Overisel, Michigan, 21.3.1884, † Cambridge, Massachusetts, 12.11.1944. George Birkhoff first studied at Harvard, where he obtained his B.A. in 1905 and his M.A. in 1906. He then went to Chicago and was granted a Ph.D. in 1907. For the following two years, he taught as an assistant professor at the University of Wisconsin, in Madison. In 1909 he became professor at Princeton. In 1912 he returned to Harvard, where he was appointed full professor of mathematics in 1919 and Perkins Professor in 1932. From 1935 to 1939, he held the position of dean of the Faculty of Arts and Science. In 1925 he was president of the American Mathematical Society, and in 1937 he became president of the American Association for the Advancement of Science. Maxime Bôcher, in Harvard, and Eliakim Hastings Moore, in Chicago, were the academic teachers who had

FIGURE 1.6   George David Birkhoff

the greatest impact on Birkhoff. He was, in addition, strongly influenced by Henri Poincaré, whose work Birkhoff very closely studied.

Birkhoff achieved world-wide recognition for the first time in 1913 when he proved Poincaré's "last geometric problem," not to be confused with the "Poincaré conjecture," which today is still unsolved. This theorem was of great significance to the three-body problem. It came to be known as the "Poincaré–Birkhoff fixed-point theorem" and today is used ubiquitously. In 1931 Birkhoff proved his "ergodic theorem," which was to become of utmost importance in the theory of dynamical systems, probability theory, group theory, functional analysis, and even in technical matters. In addition, Birkhoff designed a special relativity theory and even concerned himself with questions regarding aesthetics.

He received much recognition both in America and in Europe for his outstanding mathematical achievements. He was a member of the National Academy of Sciences, the American Academy of Arts and Sciences, and the American Philosophical Society, as well as various Latin American and European scientific associations[18] [TOEPELL 1991].

---

[18]For example, the Deutsche Mathematiker-Vereinigung.

Because of his creativity and versatility, Birkhoff had, as a teacher and as a researcher, a great impact on his numerous students. He was one of the most important American mathematicians at the beginning of the 20th century. The expression that American mathematics has grown out of its (European) infancy and has become self-sufficient is due to George Birkhoff.

For the first little while after Birkhoff, not much in connection with the Four-Color Problem was accomplished. Small incremental steps towards mastering it were taken by Philip Franklin, Alfred Errera, C.N. Reynolds, C.E. Winn, Chaim Chojnacki-Hanani, Henri Lebesgue, and Arthur Bernhart.

It wasn't until the early 1960s that the mathematician Heinrich Heesch finally succeeded in making a decisive breakthrough. He systematized the proof of reducibility and developed an algorithm for the proof of "D-reducibility," a notion he himself created and which lent itself to implementation with the aid of computers. In addition, he invented the construction of "unavoidable sets of configurations," the so-called discharging procedures. In 1964, for the actual adaptation of his algorithm to computer methods (that is, the programming), Heesch secured, with Wolfgang Händler's encouragement, the help of Karl Dürre, who was a secondary school teacher at the time.

By today's standards, Dürre practically had to work with prehistoric methods. He wrote the program in Algol 60 and punched the appropriate data (which consisted essentially of adjacency matrices) onto punch cards. A short extract of his original program is printed on page 205.

The first test run took place on November 23, 1965, on a CDC 1604A computer at the computing center of the Institute of Technology, in Hannover. This test confirmed the reducibility of the Birkhoff diamond, a configuration already known to be reducible. In December 1965, for the very first time, a configuration of ring size 9 whose reducibility had been theretofore unknown was tested for reducibility with the aid of the computer. At that time, due to limitations of computer memory modules, only figures having ring size less than 12 could be tested for reducibility. With the development of "ranking methods" for sets of boundary colors, it was then possible, from 1967 onwards, to deal only with the characteristic functions of these sets. In this way, the memory requirements were considerably lessened. A description of these methods was the subject of Dürre's dissertation [DÜRRE 1969].

In the United States, computer technology was already at a further stage of development. It was there that Heesch, indirectly through Wolfgang Haken, was drawn to the attention of Yoshio Shimamoto, the chairman of the applied mathematics department at the Brookhaven National Institute Laboratory, in Upton, New York. Shimamoto had already long toyed with the Four-Color Problem himself. In 1967 he invited Heesch to Brookhaven and generously offered him computing time on their CDC 6600.

In the following years, Heesch went, on two separate occasions, to the United States for extended visits in order to advance his work (at first with Dürre's help) on the big computer there. For that, Dürre had, first of all, to transcribe his program from Algol into Fortran.

In the end, Heesch, after yet another short stay in the United States, returned to Hannover in the hope of solving the Four-Color Problem by himself in Germany. However, to Heesch's bitter disappointment, the German research community, because certain experts had negative opinions about the project, did not allocate resources to him for the necessary large amounts of computing time and for personal costs.

A special talent of Heesch's lay somewhat outside of rigorous mathematics. It was in the "visualization" of figures created by systematically restricting graphs to vertex colorings and then observing optical features of the degrees of vertices. Through this technique, it became possible, or at least was made essentially easier, to visualize and distinguish on a somewhat grander scale. The "pared images" were fundamentally utilized for the organization and the presentation of the almost 2000 figures that were required for the proof of the Four-Color Theorem.

Heesch, Heinrich

* Kiel 25.6.1906, † Hannover 26.7.1995.
From 1925 to 1928, Heesch studied mathematics and physics in Munich,

FIGURE 1.7 Heinrich Heesch

primarily with Constantin Carathéodory and Arnold Sommerfeld. Concurrently, he studied violin with Felix Berber at the University of Music. Ultimately, after graduating from both disciplines, he decided to pursue mathematics and obtained his Ph.D. at the University of Zurich with two papers in structure theory [HEESCH 1930]. Subsequent to that, he went to Göttingen as an assistant to Hermann Weyl. In the following years, he produced significant mathematical results in new and leading research in crystal geometry. He also solved the tiling problem (Hilbert's Problem 18, see [MILNOR 1976]), which had been put forward by Hilbert in 1900 at the second International Mathematical Congress.

The rise to power of the National Socialists, however, set him back somewhat, for Heesch was forced to leave the university. For the next twenty years, he worked as an independent scholar, as a teacher of music and mathematics in various schools, and as an advisor to industry in connection with research on tilings in the plane. The findings of this kind of research, especially during the war years, were of great interest because of their possible applications to the economy of use of materials.

It was only in 1955 that Heesch again began teaching duties at the Institute of Technology, in Hannover, where he received his certificate of habilitation[19] *in 1958. Finally, in 1966, an associate professorship was conferred upon him.*

Heesch had turned his attention to the Four-Color Problem towards the end of the 1940s, a fact that is corroborated by his mathematical works from that time. He presented his significant results in many lectures but delayed their publication until 1969, when his foundational book, which had been developed out of his habilitation thesis, first appeared [HEESCH 1969]. *Other important research results remained for the most part as preprints, as did the work on E-reduction* [HEESCH 1974]. *For that reason, Heesch's name and research were not even mentioned in Ore's book, which is internationally regarded as an authoritative source* [ORE 1967].

Hans-Günther Bigalke has depicted Heesch's almost tragic role in the history of the Four-Color Problem in a full-length, if not completely objective, biography [SCHREIBER 1989] [SIEGMUND-SCHULTZE 1990]. For their part, Appel and Haken have devoted a good amount of space to Heesch's research in an article outlining the history of the Four-Color Problem, which was published in the "Mathematics Today" collection [APPEL and HAKEN 1978].

Dürre, Karl P.

* *Reichenbach/Oberlausitz 20.7.1937.*
Karl Dürre studied mathematics and physics in Marburg and Hannover and passed the state exams in 1963. During his course of studies, he found coloring problems interesting and, because of this, came to be acquainted with Heinrich Heesch. When he graduated, he became Heesch's collaborator.

He received his doctorate in January 1969 at the Institute of Technology, in Hannover. His thesis dealt with the study of sets of signs [DÜRRE 1969].

In the latter part of 1969, after his return from Brookhaven, he directed his interests more towards computer science. In the beginning, he

---

[19] In German universities, the *Habilitation* is an academic degree after the Ph.D., usually a precondition for becoming professor at a university.

FIGURE 1.8 Karl P. Dürre

was an assistant at the University of Erlangen-Nürnberg, and later, from 1971 on, he held a research position as "Wissenschaftlicher Mitarbeiter"[20] at the Institute of Technology, in Karlsruhe. After that, he worked only sporadically on the Four-Color Problem.

Since 1970 his works have focused on two main issues: firstly, on the development and analysis of algorithms and data structures (including graph-coloring algorithms and their application to the compression of sparsely populated data bases) and, secondly, on the development of user-friendly programming systems.

In particular, he was instrumental in developing pioneering techniques that considerably facilitated blind persons in accessing computer technology. One successful program to aid blind students at the University of Karlsruhe was initiated and designed by him. It was implemented shortly before his departure to the United States, where he has been, since 1987, associate professor of computer science at Colorado State University, in Fort Collins, until his retirement in 1994.

---

[20] A temporary position as a research scientist.

In terms of the work involving the recognition of reducible configurations and the development of discharging procedures, Jean Mayer has also claimed the spotlight, next to Heesch, of course. He was another outsider to the world of mathematics. He was a humanist—a literature professor—to whom we have already referred in connection with the Heawood problem.

Despite the fact that Mayer was an outsider, or maybe precisely because of it, he had brilliant ideas, which led to the aforementioned results. His raising the Birkhoff number to 96 would be subsequently surpassed by the complete proof of the Four-Color Theorem of Appel, Haken, and Koch and would, therefore, not be published.

The very efficient discharging procedures that Mayer, independently from Heesch, developed led to a collaboration with Appel and Haken [APPEL, HAKEN and MAYER 1979] and ultimately were incorporated into the proof of the Four-Color Theorem.

It is worth mentioning that Mayer, too, was able to establish a connection between his area of expertise, French literature, and his hobby, the Four-Color Problem. That happened in the following way.

In the diaries of the French lyricist Paul Valéry, in the year 1902, Mayer discovered twelve pages containing substantial notes on the Four-Color Problem. He studied Valéry's ideas and determined that certain figures, which much later would be noticed by Birkhoff, Franklin, Winn, Chojnacki-Hanani, and others, had already been found by Valéry [MAYER 1980a,b].

Mayer, Jean

∗ Champagnole/Jura 26.6.1925.
Jean Mayer studied classical languages at the University of Rennes and obtained his doctorate in 1960 at the Sorbonne with a thesis entitled "Diderot, Man of Science." From 1960 to 1965, he was professor at the Centre d'Enseignement Supérieur (Center for University Teaching), in Abidjan, the capital city of the Ivory Coast. Subsequent to that, until his retirement in 1988, he was professor at the Paul Valéry University, in Montpellier.

Mayer was coeditor of the collected works of Diderot and Voltaire. His boyhood love of mathematics led him time and again to read mathemat-

FIGURE 1.9   Jean Mayer

ical works. Berge's Theory of Graphs and Their Applications [BERGE 1958] *particularly impressed him. Its clear and engaging presentation inspired him to work independently on coloring problems. In 1966 he began to work in this area in earnest. He discovered, as is already mentioned above, the notes of Paul Valéry, and he successfully occupied himself working with discharging procedures, a topic that became important to the solution of the Four-Color Problem.*

The time of Heesch's visits to America signaled in some way the final stretch on the road to the solution of the Four-Color Problem. In the first place, it began an extended contact between Heesch and Haken, who became increasingly interested in the problem. Written communications and oral discussions between them centered essentially on Heesch's successes with the problem and around possible methods of solving it.

In the second place, in 1971, through Haken, Kenneth Appel became interested in the problem, and they began to work together on it. Up to this time, scarcely any hope existed for being able to solve the problem using the qualitative method. Tutte, *"arguably the finest graph theorist of our time"* (Appel), expressed the opinion that only an optimist could surmise the existence of an unavoidable set of reducible configurations with "only" 8000 elements [TUTTE 1974]. He therefore gave more credibility to the quantitative approach [TUTTE 1975] and clearly stated his uneasiness about computer-based solutions, such as those being put forward by Shimamoto [WHITNEY and TUTTE 1972].

Up to this point, Heesch's work had been published only in German. The last aforementioned work of Tutte and Whitney made Heesch's ideas known to the English-speaking world, and thereby a flood of various contributions to the Four-Color Problem was released. Authors of interesting works include Frank Allaire, from the University of Manitoba (Canada), and Edward Reinier Swart, from the University of Rhodesia, who worked together as guests of Tutte in Waterloo; Frank R. Bernhart, from Shippensburg State College, Arthur Bernhart's son, who made a name for himself as an expert in uncovering errors in "proofs" of the Four-Color Theorem; Thomas Osgood, doctoral student of Haken; and Walter R. Stromquist, from Harvard University.

In their joint work, Appel and Haken studied sets having up to one million elements. They placed their hopes on making rapid progress using electronic computers. Thus, in their work they emphasized the development of more powerful discharging procedures, which to implement would also require the aid of computers for more than just reduction calculations (which had hitherto been the case).[21]

It was Appel who was the programmer. He wrote a program that was capable of contributing to its own improvement.[22] In the latter part of 1975, the resulting calculations led ultimately to a surprise, that an unavoidable set out of only 2000 reducible configurations

---

[21] A complete presentation of Haken's ideas and his preliminary methods, as well as his joint work with Appel, can be found in Donald MacKenzie's survey article, "Slaying the Kraken. The Sozio-History of a Mathematical Proof" [MACKENZIE 1997]. In the article, the author, a sociologist holding a chair in sociology at the University of Edinburgh, presents the Four-Color Problem and its history to a lay readership. Towards the end, he debates, in particular, the (philosophical) question as to whether the solution given by Appel and Haken demands a revised definition of the notion of a "proof."

[22] It therefore involved a CAD technique (CAD meaning computer-aided design).

FIGURE 1.10   Picture of Envelope

was now within the realm of possibility. During a stroll on the beach while on vacation in Florida in December 1975, Haken had the brilliant idea of a new discharging procedure—the "transversal" discharging. This was, however, not as easily adaptable to computer methods. Appel and Haken then turned back to more conventional hand calculations. Simultaneously, they obtained a new and third member of the team—a graduate student named John Koch. He took over the reducibility calculations that up to that time had been neglected.

With the help of an IBM 360 in Urbana (Illinois), the complete proof of the Four-Color Theorem was finally achieved in 1976.

In 1986, in the *Mathematical Intelligencer*, Appel and Haken gave a short but crystal-clear overview of how the proof worked and also responded to the arguments against it, that had been previously put forward [APPEL and HAKEN 1986]. In the course of the proof, thousands of figures were discussed, and naturally some errors were made. These errors were divided into three categories. In 1981, in his Diplom[23] thesis, a student of electrotechnology at the Aachen Institute of Technology, Ulrich Schmidt, went over about 40% of the "proofs involving unavoidableness."[24]

He discovered close to fourteen trivial mistakes and also an error of category three. On the opposite side of the world, in Tokyo, a Japanese student named Shinichi Saeki similarly examined some of the proofs and also discovered a number of mistakes.

In the meantime, Appel and Haken had developed an "error-correction routine" that up to that time had always worked successfully. That was a characteristic feature of their proof: *When an interim step such as an error became apparent, then the whole deck didn't collapse—as is frequently the case in other mathematical proofs. It didn't become an insurmountable obstacle—only a small glitch that one could relatively easily circumnavigate.* Haken described the situation in the following statement: *"It was not a question of a* single *proof for the Four-Color Theorem, but of* many."

All the known errors up to 1989 have been addressed and corrected by Appel and Haken in a revised version of their proof [APPEL and HAKEN 1989].

Haken, Wolfgang

* Berlin 21.6.1928.
Wolfgang Haken studied mathematics, physics, and philosophy in Kiel and obtained his doctorate in 1953, specializing in topology under the supervision of Karl Heinrich Weise. Already as a young student, he was made aware of the Four-Color Problem. In 1948 he attended a lecture given by Heesch in which Heesch presented some of his first deliberations and results.

From 1954 until 1962, Haken worked in Munich as an engineer in the development of microwave technology for the company Siemens & Halske.

---

[23]The Diplom degree plays the same role in German universities as the Master of Science degree does in North America.
[24]The fact that only 40% were examined was due to the time limitation set down in the examinations' regulations for Diplom theses.

FIGURE 1.11   Wolfgang Haken

In 1962, after his discovery of a finite procedure for deciding whether a knot is knotted or unknotted (an old and classical problem that had stood hopelessly open for a long time), he was invited as visiting professor to the University of Illinois, in Urbana. From 1963 to 1965, he was a temporary member of the Institute for Advanced Study, in Princeton. Finally, in 1965, he became professor at the University of Illinois.

In 1990, the University of Illinois gave Haken the highest honor granted for scientific achievement. They nominated him for membership in the Center for Advanced Study. At the time, only fifteen scientists belonged to this center.

In the summer of 1993, the Johann Wolfgang Goethe University, in Frankfurt (Main) conferred an honorary doctorate on Haken in recognition of his work—primarily for his work on three-dimensional manifolds, which are to be named Haken manifolds after him.

Appel, Kenneth

* *Brooklyn, N.Y. 8.10.1932.*
Appel earned his bachelor of science degree from Queen's College of the City University of New York in 1953. In 1955 he received his master's

degree from the University of Michigan, where, four years later, he also obtained his doctorate under Roger Lyndon. His thesis was entitled "Two Problems on the Borderline of Logic and Algebra."

Over the years, his primary research interests have varied from the theory of recursive functions to combinatorial group theory to combinatorics. In 1956, he became interested in computers and since then has written many programs in various computer languages for mathematical and nonmathematical applications.

He became full professor at the University of Illinois, in Urbana, in 1961. In 1971, when Haken (who had also taught in Illinois since 1965) held a series of seminars concerning Shimamoto's work, Appel began in earnest to focus his attention on the Four-Color Problem. His contribution to the solution of the problem lay mainly in the programming of the discharging procedures and in the ensuing impetus to tackle the Four-Color Problem itself using these procedures.

Since the Autumn of 1993, Appel has been the chairman of the mathematics department at the University of New Hampshire.

FIGURE 1.12   Kenneth Appel

Koch, John

* Urbana, Illinois 31.10.1948.

John Koch studied at Bucknell University and at the University of Illinois, in Urbana. It was there, through Appel and Haken, that he became acquainted with the Four-Color Problem, and it was there that he wrote the computer program leading to its solution. His dissertation, entitled "Computation of Four Color Irreducibility" [KOCH 1976], evolved out of his work with the problem, and in 1976 he obtained his Ph.D. Subsequent to that, he went to Wilkes University, in Wilkes-Barre, Pennsylvania, where he held various positions as a research scientist and where since 1989 he has been full professor.

From 1983 to 1988 he was responsible for a five-year program that was to promote greater utilization of computers outside of the university. Between 1978 and 1989, he spent several summers at the Institute for Defense Analyses, in Princeton, where he took part in the development and execution of many programs on the Cray 1 and the Cray 2 computers.

In 1989, in appreciation of Kempe's contribution, Appel and Haken declared:

FIGURE 1.13 John Koch

> *Kempe's argument was extremely clever, and although his "proof"*
> *turned out not to be complete, it contained most of the basic ideas*
> *that eventually led to the correct proof one century later.*

That Kempe's result was indeed a good starting point for the proof cannot be denied. However, it must be acknowledged that Birkhoff, Heesch, Appel, and Haken still contributed very crucial ideas towards the proof.

Francis Guthrie's little bit of graphical play was to be the instigator of a century-long mathematical story in which the tree of mathematical knowledge would produce many new stems and branches, one being the theory of graphs. Nonetheless, Appel, Haken, and Koch's submitted proof of the Four-Color Theorem has not ultimately put an end to the Four-Color Problem. Many mathematicians are attempting to construct a proof independent of the use of computers, and large numbers of unanswered questions are arising out of this. In the meantime, a more metamathematical discussion over the significance of computer-assisted proofs has ensued. Such proofs have formed the basis for this book and have obtained a fresh impetus through the interim success in providing evidence of the nonexistence of an affine plane of order 10 using a Cray supercomputer. We will close the historical chapter with an almost poetic quotation from Tutte that pointed to farther-reaching and deeper future research:

> *For the Four Color Problem is just one member, one special case,*
> *of a great association. It was singled out for mathematical attack*
> *because it seemed likely to be the easiest member. But now the*
> *time has come to confront the other members of the family. . . .*
> *The Four Color Theorem is the tip of the iceberg, the thin end of*
> *the wedge and the first cuckoo of Spring* [TUTTE 1978].

From the time that the German edition of this book was first published, in January 1994, until now, the situation has changed. It sometimes happens in mathematics that an outstanding conjecture has been proven to be true, but the proof is rigid and complicated. At a later date, other mathematicians, knowing that the theorem is valid, have been able to provide a much simpler version of the proof. Good examples of this phenomenon are the proofs that Euler's number e (Hermite 1873) and Ludolph's number π (Lindemann 1882) are transcendental numbers. In most modern textbooks, the proofs are developed along the lines set out by David Hilbert in 1893. The same thing has happened to the Four-Color Theorem.

In August 1994, at the International Congress of Mathematicians, in Zürich, Paul D. Seymour, from the Bellcore Company, in Morristown (New Jersey), presented a simplified proof of the Four-Color Theorem [SEYMOUR 1995]. He, together with Neil Robertson, of Ohio State University, in Columbus, as well as Daniel P. Sanders and Robin Thomas from the Georgia Institute of Technology, in Atlanta, had worked together on its formulation. The principal ideas behind their proof will be sketched in the latter parts of Chapters 6 and 7. They, too, could not dispense with the need for computer involvement. However, they were successful in reducing the amount of computing time to a tolerable level. Those who have the presented programs at their disposal and who have understood the theoretical foundations can, within half a day, duplicate the proof. However, the question of a computer-free proof still remains!

2

(Topological) Maps

The development of the Four-Color Theorem from a mathematical perspective requires rigorous definitions of its concepts. To organize these ideas is not a simple task, and it requires a certain amount of effort. Inevitably, this will lead to some very abstract thinking. However, initially, we shall be guided by our intuition. Out of this intuition, geometric–topological objects called maps will be realized. Although the fundamental ideas arise in a set-theoretic topological setting, the Four-Color Theorem itself is of a combinatorial nature.

In this chapter the intuitive idea of a geographical map, together with its structural elements, such as its borderlines and its multinational corners, will be defined in a topological sense. They will then be converted to an abstract format from which the transition to the combinatorial setting is made possible.

2.1 Preliminary Heuristic Considerations

For the moment, we will make do with an imprecise manner of speaking because we still do not have a grasp on precise concepts.

43

We will say that we have an *admissible coloring* of a map, or, in other words, a map that is *admissibly colored*, if each country of this map is colored in such a way that any two countries having a common borderline are colored with different colors. An *admissible n-coloring*, where *n* denotes a natural number, is an admissible coloring using at most a total of *n* colors. Naturally, in the case of the Four-Color Theorem, we are primarily interested in admissible 4-colorings of a map.

By a map, we mean a partition of an (infinite) plane into finitely many countries that are divided from one another by borders and of which only one country is *unbounded*. A set of points in the plane is said to be *unbounded* if no rectangle completely encompasses it. A set is called *bounded* if it is entirely contained within some rectangle.

There is no essential difference between the notion of an infinite plane with one unbounded country and an actual geographical map. A map in an atlas is, in general, contained within a rectangular region. We consider the portion of the plane exterior to that rectangular region as another country. Therefore, a successful admissible *n*-coloring of the entire plane is also an admissible *n*-coloring of the rectangular map. Conversely, if we know that we can color every rectangular map admissibly with *n* colors, then we can obtain admissible colorings for maps covering the entire plane. Since we assume that only one unbounded country exists, we are always able to find a rectangular region whose perimeter lies entirely within the unbounded portion. An admissible *n*-coloring of a rectangular map also supplies a color for the part of the unbounded country that is contained within the rectangular region. This color can then be used as a color for the entire unbounded country.

At this juncture we would like to mention that by partitioning the globe into various countries, we can similarly talk about an *admissible n-coloring* of a map of the world. Moreover, the problem of the existence of an admissible 4-coloring of the globe is, in fact, equivalent to the problem of the existence of such a coloring for the plane. One can imagine a globe representing the earth as a spherical balloon out of which a tiny hole in the interior of some country has been extracted. The remaining portion can then be stretched over a rectangular map. Therefore, if one has such a coloring for each plane map, then each map of the globe can be admissibly colored with *n* colors. Conversely, if an admissible n-coloring exists for each map on a globe and one has a map on a plane, then one can wrap this plane map onto the punctured globe so that the hole is placed inside the map's unique unbounded country. One can now color the map that has been wrapped around the globe, and hence the original plane map, admissibly with n-colors. This intuitive transition between the sphere and the plane, a topic that is frequently debated in connection with the Four-Color Theorem, will be accomplished formally by using the "stereographic projection" [DO CARMO 1976, page 67].

The condition "only *one* unbounded country" actually presents no real limitation. If we have several unbounded countries, then we also have borderlines extending to infinity. We wrap this map around the sphere so that the hole consists of exactly one point. We collapse the previously infinite borderlines onto this point and thus obtain a partition of the entire sphere. If we are able to color this partition admissibly with n colors, then it is also evidently possible to do so for the original plane map.

There is one essential restriction, however, which is meaningful and even necessary. We will be considering only maps having countries that are connected. Consequently, we do not allow a country to consist of distinct separated regions, such as the United States of America. It is sometimes possible to color admissibly with 4 colors certain maps having countries with distinct disconnected regions. However, here is one example for which no such coloring exists:

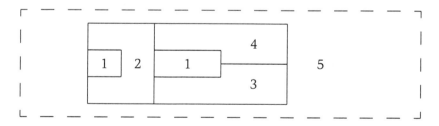

The above map has five countries, which are indicated by the numbers 1 to 5. We will now attempt to color it. We observe that each of the countries from 1 to 4 has common borderlines with the other three. Thus we require four different colors for these countries. Country number 5, however, has common borderlines with all of the other countries. Hence, none of the previous colors can be utilized to color country 5.[1]

From the above map, we are able to obtain a map with only connected countries if we combine the western portion of country 1 with country 2. Then it still is the case that each of countries 1 through 4 abuts the other three. Consequently, at least four colors are again necessary. However, country 5 can be colored with the same color as country 1. This also shows that in the event that all countries are connected, fewer than four colors is not possible in general.

2.2   Borderlines

A map is created by the drawing of its borders. These borders are lines. The points at which these lines meet may be specially marked. If so, we call these points "multinational corners." Therefore, we must first make precise the notion of a (border)line. This will be done in this section, where we will discuss at length the geometric–topological foundations and in which we will already derive a few of the combinatorial results. The topological concepts will be explained in so far as is necessary for an understanding of the stated results. For the proofs, however, we assume a basic knowledge of topology. On the first reading, the proofs can be skipped if one can make the correlation between the expressed claims and the appropriate diagrams. Due to the limited space in this book, we must forgo complete proofs of some of the results—in particular, of two of the very deep fundamental theorems on the topological nature of the Four-Color Theorem. These are the Jordan curve theorem (Theorem 2.2.5) and the Schoenflies theorem (Theorem 2.2.7).

Intuitively, we imagine a borderline as a set of points in the plane that one can traverse in a unit of time ("unit interval") and that has no breaks ("continuous"), no stationary points and no self-intersections

---

[1] Heawood dealt comprehensively with the coloring of maps having countries with regions disconnected from one another [Heawood 1890].

("injective"). We distinguish two cases by specifying whether or not the initial and terminal points of the interval coincide. In the following definition, $[0, 1]$ denotes the *(closed) unit interval*, that is, the set of all real numbers t such that $0 \le t \le 1$. The *right half-open unit interval*, denoted by $[0, 1)$, consists of the set of all real numbers t where $0 \le t < 1$. The set of all real numbers will be designated by $\mathbb{R}$, the set of all points in the plane by $\mathbb{R}^2$. This means that we imagine the plane endowed with a coordinate system in which each point in the plane is identified with the real number pair $\mathbf{x} = (x_1, x_2)$, considered to be its coordinates. A mapping $f : [0, 1] \to \mathbb{R}^2$ is said to be *continuous* if its *coordinate functions* $t \mapsto f_1(t)$, $t \mapsto f_2(t)$, for $t \in [0, 1]$, are continuous. The functions $t \mapsto f_1(t)$ and $t \mapsto f_2(t)$ are, respectively, the first and the second coordinates of the point $f(t)$. A mapping $f : [0, 1] \to \mathbb{R}^2$ is defined to be *injective* if distinct points of the interval are assigned distinct function values. In other words, if $t_1 \ne t_2$, then $f(t_1) \ne f(t_2)$.

Definition 2.2.1
A subset C of the plane $\mathbb{R}^2$ is

(a) an *arc* if there exists an injective continuous mapping
$c : [0, 1] \to \mathbb{R}^2$ such that

$$C = \text{Image}\, c := \{\, c(t) : t \in [0, 1] \,\};$$

(b) a *simple closed curve*, or a *closed Jordan curve*, if there exists a continuous mapping $c : [0, 1] \to \mathbb{R}^2$ for which

$$C = \text{Image}\, c$$

and such that $c(0) = c(1)$ and $c|_{[0,1)}$ is injective;

(c) a *simple curve* if it is either an arc or a closed Jordan curve.

The name Jordan is in honor of Camille Jordan, who proved the aforementioned Jordan curve theorem (Theorem 2.2.5) in the years 1887–1893. If C is a simple curve, then a mapping c having the properties given in Definition 2.2.1 is called a *parametrization* of C. In what follows, we will use the notation outlined below:

C, with or without indices, with a bar over it, or otherwise adorned, for arbitrary simple curves.

B, with or without embellishment, for arcs.

K, with or without embellishment, for closed Jordan curves (these are deformed circles).

The above definitions, although intuitive, are very general. They include, for example, the nowhere smooth (meaning nowhere differentiable) "snowflake curve," which was defined by Helge von Koch in 1904 [CUNDY and ROLLET 1961, Section 2.10.1, page 65]. They also include William Fogg Osgood's curve, from 1903, which has a positive surface area [GELBAUM and OLMSTED 1964, Chapter 10, Example 10], in contradiction to Euclid's definition "Γραμμὴ δὲ μῆκος ἀπλατές" [A line is a length having no width, EUCLID 1883/1969, Volume I, Definition 2]. The whole idea of what a curve is constitutes a true "house of horrors" [HEUSER 1986, page 367]. For the problem of the coloring of maps, however, the concepts we have just defined suffice. This is because on the one hand, the requirement of injectivity precludes the surface-covering "Peano curves" [GELBAUM and OLMSTED 1964, Chapter 10, Example 6], for instance an entire square-filling curve.[2] On the other hand, it is because of the following two facts which emphasize what a line-like object is.

From Corollaries 2.2.8 and 2.3.12, every point $\mathbf{x}$ of a simple curve C is arcwise accessible from within $\mathbb{R}^2 \setminus C$, the complement of C, that is the set of all points in the plane not belonging to C. We define a point $\mathbf{x}$ of a simple curve C to be *arcwise accessible from within a set M* if there exists an arc from $\mathbf{x}$ to a point of M that, except for $\mathbf{x}$, lies entirely within the set M [MOISE 1977, page 70].

A simple curve is a nowhere dense subset of the plane (Theorem 2.3.14). Note that an arbitrary set M of the plane is said to be *nowhere dense* if for each disk in the plane there exists a (possibly smaller) disk containing no points of M.

To clarify these properties, we consider the Peano curve (a stylized snake that is in the process of digesting a rabbit) shown below.

---

[2]This curve was named after Giuseppe Peano, the originator of the well-known axiom system for the natural numbers.

Note that the center of the completely shaded disk is not arcwise accessible from within the complement of the curve and that every disk that is completely contained in this shaded disk, has a nonempty intersection with the curve. In this case, we have a curve that is not a simple curve. The specified properties imply that with simple curves, thickenings are precluded.

It must, however, be emphasized that many seemingly almost self-evident statements and theorems are sometimes very difficult to prove rigorously.[3] For this reason, we must—as has already been mentioned at the beginning of this section—dispense with a few of the topological proofs. For these proofs, however, we will give precise references.

In any case, we still need a few topological ideas. Let M be a subset of the plane. A point $\mathbf{x} \in \mathbb{R}^2$ is defined to be a *boundary point* of M if each disk centered at $\mathbf{x}$ contains both points of M and points not in M. A boundary point of M may or may not belong to M. The set of all boundary points of a set M is called the *boundary* of M and will be denoted by $B(M)$. A point $\mathbf{x}$ is said to be an *interior point* of M if it belongs to M but is not a boundary point of M. In this case, we say that M is a *neighborhood* of $\mathbf{x}$. The set M is said to be *open* (in $\mathbb{R}^2$) if it contains none of its boundary points. An open set M, therefore, contains only interior points. In other words, for each $\mathbf{x} \in M$, there exists a disk centered at $\mathbf{x}$ that is entirely contained in M. The set M is defined to be *closed* in $\mathbb{R}^2$ if it contains all of its boundary points. This is precisely the case when $\mathbb{R}^2 \setminus M$, the complement of M, is an open set. It should be noted, firstly, that a set of the plane is not necessarily either open or closed. Secondly, observe that as subsets of the plane, the empty set and $\mathbb{R}^2$ itself are both open and closed. Finally, a set is called *compact* if it is closed and bounded. We can now state the first important property of simple curves.

Lemma 2.2.2
Every simple curve is a compact set.   ■

Proof See [Dugundji 1966, Chapter XI, Theorem 1.4(i)].   □

---

[3]A fitting remark related to this observation can be found in [Moise 1977, page 78]: "Some reflection may be needed to convince oneself that this theorem is not trivial."

Geometric objects are considered topologically the same if one can be deformed into the other by bending, stretching, or shrinking—but without gluing or tearing. From this observation comes the notion of a mapping between two geometric objects being a *homeomorphism* if it is bijective and if the mapping and its inverse are both continuous. Recall that a mapping f between geometric objects is said to be *continuous* if for all a in its domain the following holds:[4]

$$\lim_{x \to a} f(x) = f(a).$$

Thus two geometric objects are said to be *homeomorphic* if there exists a homeomorphism between them. In this context, first of all, the following holds:

Lemma 2.2.3
Every arc is homeomorphic to the closed unit interval $[0, 1]$.   ∎

Proof   See [DUGUNDJI 1966, Chapter XI, Theorem 2.1(2)].   □

A special case of a simple closed curve (closed Jordan curve) is the unit circle

$$S^1 := \{ (x, y) \in \mathbb{R}^2 : x^2 + y^2 = 1 \}.$$

The mapping

$$k_S : [0, 1] \to \mathbb{R}^2, \quad t \mapsto (\cos 2\pi t, \sin 2\pi t)$$

is a parametrization of S^1. The following result comes from very elementary topological ideas.

Lemma 2.2.4
Every closed Jordan curve is homeomorphic to the unit circle S^1.   ∎

Proof   Let K be a closed Jordan curve, and let $k : [0, 1] \to \mathbb{R}^2$ be a parametrization of K. Then the mapping $h : S^1 \to K$, $k_S(t) \mapsto k(t)$, is continuous. This is because the mapping k_S is closed [DUGUNDJI 1966, Chapter XI, Theorem 2.1(1)] and because the composition $h \circ k_S' = k'$ is continuous, where $k_S' : [0, 1] \to S^1$ and $k' : [0, 1] \to K$ are the

---

[4]The interested reader who is not yet conversant with this material may wish to convince him/herself that this condition generalizes the definition for continuity of mappings $[0, 1] \to \mathbb{R}^2$ which is given on page 47.

mappings induced from k_S and k, respectively. Since the mapping h is, in addition, bijective, it follows that h is a homeomorphism [DUGUNDJI 1966, Chapter XI, Theorem 2.1(2)]. $\square$

An arc *joins* two distinct points of the plane—namely its *end points*, or its *boundary points*. Of special significance is the shortest path between two distinct points $\mathbf{x}_0$ and $\mathbf{x}_1$. This is the arc

$$[\mathbf{x}_0, \mathbf{x}_1] = \{(1-t)\mathbf{x}_0 + t\mathbf{x}_1 : t \in [0,1]\},$$

which will be called the *straight-line segment*, or simply the *line segment*, from $\mathbf{x}_0$ to $\mathbf{x}_1$. In general, an arc is called *straight* if it is the line segment joining its end points.

A closed Jordan curve *partitions* the plane in the following way:

Theorem 2.2.5 (Jordan Curve Theorem)
Let K be a closed Jordan curve. Then $\mathbb{R}^2 \setminus K$ is the disjoint union of two open sets $I(K)$ ("interior domain" of K) and $A(K)$ ("exterior domain" of K) in such a way that:
1. *$I(K)$ is bounded, but $A(K)$ is unbounded.*
2. *$I(K)$ and $A(K)$ are path connected.*
3. *Every arc that joins a point of $I(K)$ to a point of $A(K)$ has at least one point in common with K.*
4. *Every neighborhood of a point of K has a nonempty intersection with $I(K)$ and with $A(K)$.* $\blacksquare$

A set M is called *path connected* if every two distinct points of M can be joined by an arc that lies entirely in M. Properties 2 and 3 above imply that $I(K)$ and $A(K)$ are the path components of $\mathbb{R}^2 \setminus K$. A set L is defined to be a *path component* of the set M if L is a nonempty path-connected subset of M and if no element of L can be joined to an element of its complement $M \setminus L$ by a simple arc lying entirely in M.

A complete proof of the Jordan curve theorem using the so-called winding number was given by Erhard Schmidt [SCHMIDT 1923]. A more modern proof can be found in [MOISE, 1977, Chapter 4]. At this point, unfortunately, we are not able to go into this in any further detail.

Let C be a simple curve and $c : [0,1] \to \mathbb{R}^2$ a parametrization of C. If C is an arc, then the points $c(0)$ and $c(1)$ are called its *initial* and *terminal* points, respectively. In this case, we say that C is an arc *from* $c(0)$ *to* $c(1)$. If C is a closed Jordan curve, then the point $c(0) = c(1)$ is called its *emanating point*. The initial and terminal points of an arc are distinct from one another, whereas a closed Jordan curve winds back to its emanating point. It should be noted that the end points of an arc are uniquely determined, regardless of the choice of parametrization. However, depending on the choice of parametrization, either one of them may be regarded as the initial point, respectively terminal point, of the arc. *Interior* points of an arc are those points that are distinct from the end points. Be careful: By this we do not mean "interior points" in the general topological sense (see the definition on page 49). The set of interior points of the arc B will be denoted by $\overset{\circ}{B}$. No distinction is made between end points and interior points for closed Jordan curves. One can take any point as the emanating point.

If one chooses finitely many (at least two) points on a closed Jordan curve, then one has a partition of the curve into equally as many arcs. Conversely, one can concatenate arcs that have only end points in common to form either a new arc or a simple closed curve. A simple curve that consists of finitely many abutted line segments is called a *polygonal arc without self-intersections* in the case that it itself is an arc. It is called a *polygon* if it is a closed Jordan curve. At least three such line segments are needed to form a polygon. A polygon having exactly three line segments is said to be a *(rectilinear) triangle*. A polygon is called an *n-gon* if at least n points are needed for any partitioning into line segments. Because one can partition each individual line segment into arbitrarily many subsegments, in an n-gon more than n vertices can be chosen to obtain such a partition. Examples of special 4-gons are the *rectangle* and the *square.*

The reader may clarify the following observations by making appropriate sketches. The formal proofs follow almost automatically if one has the correct diagrams. In these proofs, it is frequently a question of whether or not two points of a given subset of the plane can be joined by a simple arc lying entirely within this subset. An important technical aid in that regard comes from the following simple fact.

Proposition 2.2.6

Suppose the arc B_1 joins the points $\mathbf{x}$ and $\mathbf{y}$ and that the arc B_2 joins $\mathbf{y}$ to $\mathbf{z}$. Then there exists an arc $B \subseteq B_1 \cup B_2$ joining $\mathbf{x}$ to $\mathbf{z}$. ∎

Proof The arcs B_1 and B_2 have at least one point $\mathbf{y}$ in common. Probably, however, they have several other common points as well. We now move from $\mathbf{x}$ along B_1 until we encounter our first point of B_2. From there, we continue along B_2 to $\mathbf{z}$. In this way, we obtain the desired arc. □

A vital connecting link between topology and the combinatorial problem of map-coloring is found in the following enhancement to the Jordan curve theorem, which was proved by Arthur Schoenflies in 1908.

Theorem 2.2.7 (Schoenflies Theorem)

Let K be a closed Jordan curve. Then every homeomorphism $h : K \to S^1$ can be extended to a homeomorphism $H : \mathbb{R}^2 \to \mathbb{R}^2$. ∎

Proof [MOISE 1977, Chapter 9]. □

For each closed Jordan curve K, there exist (many) such homeomorphisms $h : K \to S^1$ (Lemma 2.2.4). Every homeomorphism $H : \mathbb{R}^2 \to \mathbb{R}^2$ that extends one such h maps the interior domain $I(K)$ to the interior of the unit circle and the exterior domain $A(K)$ to the exterior of the unit circle.

This theorem has a few important consequences—for instance, the arcwise accessibility of a point on a closed Jordan curve from within the complement of the curve in the plane, in a slightly stronger sense than what the definition requires (see page 48).

Corollary 2.2.8

Suppose K is a closed Jordan curve. Then every point of K can be joined to every point in the interior domain (exterior domain) of K by an arc that except for the initial point lies entirely in the interior domain (exterior domain) of K. ∎

Proof Let $\mathbf{x} \in K$, $\mathbf{y} \in I(K)$, and $\mathbf{z} \in A(K)$ be given. We choose a homeomorphism $h : \mathbb{R}^2 \longrightarrow \mathbb{R}^2$ that maps the closed Jordan curve K onto the unit circle S^1. Let Z^i be the line segment joining $h(\mathbf{x})$ and $h(\mathbf{y})$. Then $h^{-1}(Z^i)$ is an arc that joins $\mathbf{x}$ to $\mathbf{y}$ and that except for the initial point lies entirely in $I(K)$. To find an arc from **x** to **z** that except for **x** lies entirely in $A(K)$ is a little more complicated because the line segment joining $h(\mathbf{x})$ and $h(\mathbf{z})$ may have a nonempty intersection with the interior domain of the unit circle. However, we can in any case find a polygonal arc (having no self-intersections) Z^a that joins $h(\mathbf{x})$ to $h(\mathbf{z})$ and that lies, except for $h(\mathbf{x})$, entirely in the exterior domain of the unit circle. Then $h^{-1}(Z^a)$ is the desired arc.   □

Using the same technique (or by using Proposition 2.2.6), we obtain:

Corollary 2.2.9
Every two distinct points of a closed Jordan curve K can be joined by arcs whose interior points lie entirely in the interior domain of K. They can also be joined by arcs whose interior points lie exclusively within the exterior domain of K.   ■

This fact is very helpful in the following assertions, whose somewhat lengthy proofs can be omitted on the first reading.

Corollary 2.2.10
Let K and K' be two closed Jordan curves having exactly two points $\mathbf{x}_1$ and $\mathbf{x}_2$ in common. Denote by B_l and B_r the arcs into which K is subdivided by the two given points. Denote by B_l' and B_r' the arcs subdividing K'. If one of the arcs B_l' or B_r' (excluding the end points) lies in the interior domain and the other in the exterior domain of K, then one of the arcs B_l or B_r lies entirely in the interior domain and the other in the exterior domain of K'.   ■

Proof   We choose points $\mathbf{y}_l \in \overset{\circ}{B}_l'$ and $\mathbf{y}_r \in \overset{\circ}{B}_r'$. The previous corollary implies that we can also choose arcs B'^i and B'^a joining these points such that $\overset{\circ}{B}'^i \subset I(K')$ and $\overset{\circ}{B}'^a \subset A(K')$. By part 3 of the Jordan curve theorem (Theorem 2.2.5), the arcs B'^i and B'^a must both intersect the closed Jordan curve K. Pick $\mathbf{z}^i \in B'^i \cap K$ and $\mathbf{z}^a \in B'^a \cap K$. The point $\mathbf{z}^i$ is an interior point of either B_l or B_r. Without loss of generality, we can suppose that $\mathbf{z}^i \in B_l$. All other interior points of B_l can be joined to $\mathbf{z}^i$ by an arc that does not meet K', namely by an arc segment of B_l. From that, it follows that $\overset{\circ}{B}_l \subset I(K')$ (again by part 3 of the Jordan curve theorem). Because $\mathbf{z}^a$ lies in the exterior domain of

K', it follows that $\mathbf{z}^a$ must be an interior point of B_r. Using a similar argument, one can show that $\overset{\circ}{B}_r \subset A(K')$.   □

We notice that the assumption "one of the arcs lies ..." is truly necessary. The conclusion would be false if, for instance, the interior points of B'_l and B'_r would all lie in the interior domain of K.

The next result is really just a corollary of the previous one.

Theorem 2.2.11

In $\mathbb{R}^2$, consider two distinct points $\mathbf{x}$ and $\mathbf{y}$ that are joined by three arcs B_1, B_2, and B_3 having pairwise no interior points in common. Then the following assertions hold:

1. *The interior points of exactly one of these three arcs lie in the interior domain of the closed Jordan curve formed by the other two.*
   In the case that this arc is B_2, then:
2. *The exterior domain of the closed Jordan curve $K_{13} = B_1 \cup B_3$ is the intersection of the exterior domains of the closed Jordan curves $K_{12} = B_1 \cup B_2$ and $K_{23} = B_2 \cup B_3$; and*
3. *the interior domain of K_{13} is the disjoint union of the interior domains of K_{12} and K_{23}, together with the set of interior points of B_2.*   ■

Proof   1.   As has already been stated in the second statement of the theorem, we denote by K_{ij} the closed Jordan curve formed by B_i and B_j, where $1 \leq i < j \leq 3$. In addition, we use the abbreviations $I_{ij} = I(K_{ij})$ and $A_{ij} = A(K_{ij})$. Because the arcs B_k, for $1 \leq k \leq 3$, pairwise have no interior points in common, each set $\overset{\circ}{B}_k$ lies either entirely in I_{ij} or entirely in A_{ij}, $i \neq k \neq j$. We now must establish the existence and the uniqueness of an index k such that $\overset{\circ}{B}_k \subset I_{ij}$, $i \neq k \neq j$.

To show existence, we can assume $\overset{\circ}{B}_3 \subset A_{12}$; otherwise, we can take $k = 3$. We choose an arc B_4 from $\mathbf{x}$ to $\mathbf{y}$ such that $\overset{\circ}{B}_4 \subset I_{12}$ (using Corollary 2.2.9). Then we form the closed Jordan curve $K_{34} = B_3 \cup B_4$, whose interior and exterior domains we denote by I_{34} and A_{34}, respectively. By Corollary 2.2.10, either $\overset{\circ}{B}_1 \subset I_{34}$ and $\overset{\circ}{B}_2 \subset A_{34}$, or $\overset{\circ}{B}_1 \subset A_{34}$ and $\overset{\circ}{B}_2 \subset I_{34}$. Without loss of generality, we can assume the latter possibility. We now claim that in that case, $\overset{\circ}{B}_2 \subset I_{13}$ also holds. Let $\mathbf{z}_2$ be an arbitrary point of $\overset{\circ}{B}_2$. Because the sets I_{ij} and the arcs that bound them are all bounded, the intersection of the sets A_{ij} is not empty. Choose a point $\mathbf{z}$ in the intersection $A_{13} \cap A_{34} \cap A_{12}$. It suffices to show that each arc that joins $\mathbf{z}$ to $\mathbf{z}_2$ has a point in common with K_{13}, hence with B_1 or with B_3 (part 2 of the Jordan curve theorem). Let B be such an arc. It may be that B and B_2 have a number of points in common, not just the end point $\mathbf{z}_2$. Starting from $\mathbf{z}$, let $\mathbf{z}'$ be the first point of B_2 encountered along the arc B. If $\mathbf{z}' = \mathbf{x}$ or $\mathbf{z}' = \mathbf{y}$, there is nothing more to show. Therefore, we can assume $\mathbf{z}' \in \overset{\circ}{B}_2$, and then it suffices to prove the claim for the arc segment B' of B that joins $\mathbf{z}$ to $\mathbf{z}'$. The arc B', consequently, has only the end point $\mathbf{z}'$ in common with B_2 and does not contain $\mathbf{x}$ or $\mathbf{y}$. Because $\mathbf{z} \in A_{34}$ and $\mathbf{z}' \in I_{34}$, B' must intersect $\overset{\circ}{B}_3$ or $\overset{\circ}{B}_4$ (part 3 of the Jordan curve theorem). If the first case occurs, we are finished. In the second case, we consider the subsegment B'' of B' that joins $\mathbf{z}$ to $\mathbf{z}''$, the first point of $\overset{\circ}{B}_4$ (starting out from $\mathbf{z}$) encountered along B'. Since $\mathbf{z} \in A_{12}$ and $\mathbf{z}'' \in I_{12}$, the arc B'' must contain a point of B_1 or of B_2. By the construction of B'', the latter is not possible. Therefore, B'' and hence B both contain a point of B_1. This is precisely what we had to show.

To prove uniqueness we can, on the basis of the proof of existence established above, without loss of generality assume that $\overset{\circ}{B}_2 \in I_{13}$. It remains to be shown that $\overset{\circ}{B}_1 \not\subset I_{23}$ and $\overset{\circ}{B}_3 \not\subset I_{12}$. To do this, we choose a point $\mathbf{z} \in A_{12} \cap A_{13} \cap A_{23}$. Then, for any point $\mathbf{z}_1 \in \overset{\circ}{B}_1$, we can find an arc B' joining $\mathbf{z}$ to $\mathbf{z}_1$ such that $\overset{\circ}{B}' \subset A_{13}$ (Corollary 2.2.8). Note that B' and K_{23} have no points in common. Since $\mathbf{z} \in A_{23}$, it follows that $\mathbf{z}_1 \in A_{23}$ (again from part 3 of the Jordan curve theorem). For $\overset{\circ}{B}_3$, we proceed in the same way.

2. It must be shown that under the assumption that $\overset{\circ}{B}_2 \subset I_{13}$, then

$$A_{12} \cap A_{23} = A_{13}.$$

On the basis of the Schoenflies theorem (Theorem 2.2.7), we can let $K_{13} = S^1$ and, in addition, let $\mathbf{x} = (1, 0)$ and $\mathbf{y} = (-1, 0)$. Then it is obvious that

$$A_{12} \cap A_{23} \supset A_{13}.$$

To prove the reverse inclusion, we assume that the opposite holds. Then we find a point $\mathbf{z} \in A_{12} \cap A_{23} \cap I_{13}$. Since $\overset{\circ}{B}_3 \subset A_{12}$, we can find an arc B_3' with $\overset{\circ}{B}_3' \subset I_{13} \cap A_{12}$ that joins $\mathbf{z}$ to a point $\mathbf{z}_3 \in \overset{\circ}{B}_3$. Analogously, we can find an arc B_1' joining $\mathbf{z}$ to a point $\mathbf{z}_1 \in \overset{\circ}{B}_1$ such that $\overset{\circ}{B}_1' \subset I_{13} \cap A_{23}$. Thus, in $B_1' \cup B_3'$ we also have an arc B' satisfying $\overset{\circ}{B}' \subset I_{13}$ that joins $\mathbf{z}_1$ and $\mathbf{z}_3$ and that contains a point $\mathbf{z}' \in A_{12} \cap A_{23} \cap I_{13}$. We now choose yet another arc B'' for which $\overset{\circ}{B}'' \subset A_{13}$ and that joins $\mathbf{z}_1$ to $\mathbf{z}_3$. The closed Jordan curve $K = B' \cup B''$, by the construction, contains no point of B_2. Now let B_r be the arc segment along S^1 joining $\mathbf{z}_1$ to $\mathbf{z}_3$ that contains $\mathbf{x}$. Let B_l be the complementary arc segment. From Corollary 2.2.10 it follows that one of the points $\mathbf{x}$, $\mathbf{y}$ is in $I(K')$ and the other is in $A(K')$. Consequently, the arc B_2 must intersect the closed Jordan curve K (part 3 of the Jordan curve theorem 2.2.5), whereby the desired contradiction is obtained.

3. That $I_{13} = I_{12} \cup I_{23} \cup \overset{\circ}{B}_2$ follows from part 2. It remains to be shown that the sets I_{12} and I_{23} are disjoint. Let $\mathbf{z} \in I_{12}$ be given. If $\mathbf{z}$ is also in I_{23}, then $\mathbf{z}$ can be joined to an interior point of B_3 by an arc B such that $\overset{\circ}{B} \subset I_{23} \subset I_{13}$. Hence, $B \cap K_{12} = \emptyset$ (Corollary 2.2.8). That, however, cannot happen because every arc that joins the point $\mathbf{z} \in I_{12}$ to a point in $\overset{\circ}{B}_3 \subset A_{12}$ must intersect the closed Jordan curve K_{12} (part 3 of the Jordan curve theorem 2.2.5). $\qquad\square$

Remark: Parts 2 and 3 of this theorem can also be proved directly from the Jordan curve theorem without using the Schoenflies theorem. They can then in turn be used to prove the Schoenflies theorem [Rinow 1975, Theorem 40.4]. $\diamond$

Part 3 of the theorem proven above has an interesting application. However, for that we need another definition. Let $\mathbf{x}_1$, $\mathbf{x}_2$, $\mathbf{x}_3$, and $\mathbf{x}_4$ be four distinct points of a closed Jordan curve K. The pairs $\{\mathbf{x}_1, \mathbf{x}_3\}$ and $\{\mathbf{x}_2, \mathbf{x}_4\}$ are said to be *separated in K* if the points $\mathbf{x}_2$ and $\mathbf{x}_4$ do not both lie on the same arc of the two arcs into which K is partitioned by the points $\mathbf{x}_1$ and $\mathbf{x}_3$. This means that this "separation" describes a symmetric relation on the pairs of distinct points of K.

Corollary 2.2.12

Let K be a closed Jordan curve. Let B and B' be arcs that respectively join two points of K but have no interior points in common with K or with each other. Furthermore, assume that the pairs of end points of B and B' are separated in K. If the interior points of B lie in the interior domain of K, then the interior points of B' lie in the exterior domain of K (and conversely). ∎

Proof   Let $\mathbf{x}_1$, $\mathbf{x}_3$ be the end points of B', and $\mathbf{x}_2$, $\mathbf{x}_4$ the end points of B. Denote by B_1 the arc segment along K from $\mathbf{x}_2$ to $\mathbf{x}_4$ that contains $\mathbf{x}_1$. Denote by B_3 the arc segment along K complementary to B_1. By assumption, $\mathbf{x}_3 \in B_3$. The assumption similarly implies that $\overset{\circ}{B}{}'$ must lie in a path component of $\mathbb{R}^2 \setminus (K \cup B)$.

Firstly, we consider the case when $\overset{\circ}{B} \subset I(K)$. Then $\mathbb{R}^2 \setminus (K \cup B)$ is a disjoint union of the open sets $A(K)$, $I(B_1 \cup B)$, and $I(B \cup B_3)$ (part 3 of Theorem 2.2.11). Because these sets are path connected, they are precisely the path components of $\mathbb{R}^2 \setminus (K \cup B)$. However, only the boundary of $A(K)$ contains both end points of B'. Therefore, $\overset{\circ}{B}{}' \subset A(K)$ must hold.

In the second case, when $\overset{\circ}{B} \subset A(K)$, we can, by a double application of stereographic projection, go back and use the first argument again.   □

We close this section with a few technical considerations that will be necessary for many of the proofs to come. We take a line segment S and an open set U in the plane such that $S \subset U$. By a *frame* of S in U we mean a rectangle R that has two of its sides parallel to S and that is entirely contained in U in such a way that $S \subset I(R) \subset U$.

Lemma 2.2.13

Each line segment contained in an open set has a frame in this open set. ∎

Proof   Suppose we are given a line segment S and an open set U with $S \subset U$. The following construction is a standard method in set-theoretic topology. For each point $\mathbf{x} \in S$, we choose a square $Q_\mathbf{x}$ that has $\mathbf{x}$ as its center, that has two sides parallel to S, and that together with its interior is entirely contained in U. The interiors $I(Q_\mathbf{x})$, for all $\mathbf{x} \in S$, together form an open cover of S. By the covering theorem of

Heine and Borel [FRIDY 1987, Theorem 3.4, page 35], S is contained in a finite union of these open sets. Consequently, we can find finitely many of the these constructed squares, say $Q_1, \ldots, Q_n$, such that $S \subset \bigcup_{i=1}^{n} I(Q_i)$. From the construction, we can assume that one end point of S is an interior point of Q_1 and that the other end point is an interior point of Q_n. If S is entirely contained in $I(Q_1)$ or in $I(Q_n)$, then we can take Q_1, respectively Q_n, as the frame of S. If not, then each of Q_1 and Q_n, respectively, has one side perpendicular to S that does not intersect S. We cut both of these sides with the straight lines that are parallel to S and that pass through the vertices of the smallest of all the squares Q_i for $i = 1, \ldots, n$. The resulting rectangle is the required frame of S in U. $\square$

2.3 Formal Definition of (Topological) Maps

A map is fully determined by its borderlines. Graphically, one can consider all borderlines from one multinational corner to another as arcs. Moreover, two multinational corners can be joined by several borderlines. For instance, on a geographical map, there are two 3-nation corners between Liechtenstein, Austria, and Switzerland. It can happen, however, that one country completely encompasses another, as is the case with Italy enclosing the tiny state of San Marino. In this case, we can consider its borderline as a closed Jordan curve. It is also possible that such a closed borderline touches a multinational corner. The number of possibilities that can occur makes the mathematical treatment really cumbersome, although Kempe did not seem to be bothered by it in his handling of the Four-Color Problem in 1879. To the contrary, he got around it very cleverly [KEMPE 1879c]. Naturally, our present-day notion of a simple curve was not yet available to him when he first attempted to prove the theorem, since clarification of the problematic nature of curves (see page 48) first took place around 1890. However, on an intuitive level, his ideas concerning curves were correct. At this point, we are able to achieve a considerable degree of uniformity by using a trick. We subdivide the actual borderlines with supplementary border-stones. We then take, as a basis for future deliberations, the resulting arcs that are created by this partitioning of the original borderlines. The partitioning occurs in such a way that each of the created arcs joins two of the border-stones but has none of the stones in its interior. We then proceed as follows:

- Every multinational corner is treated as a border-stone.

- If two multinational corners are linked by several borderlines, then we subdivide each of them with a border-stone.

- We partition every closed borderline with three border-stones (should multinational corners occur, two should suffice).

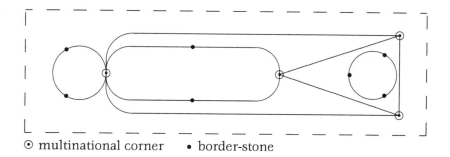

⊙ multinational corner   • border-stone

In doing this, we now have to deal only with borderlines that are arcs having the property that two distinct border-stones are linked to one another by at most a single arc. This leads us to the following abstract definition.

Definition 2.3.1
A *map* is a finite set $\mathcal{L}$ of arcs in the plane $\mathbb{R}^2$ having the property that the intersection of any two distinct arcs in $\mathcal{L}$ is either empty or is a common end point of these arcs.

This is a very simple, but noteworthy, definition. It is noteworthy because the concepts of "border-stone" and of "country" are not mentioned at all. We will talk about this again momentarily. First, however, another bit of terminology: For a given map $\mathcal{L}$, we denote the arcs belonging to $\mathcal{L}$ as *edges* of $\mathcal{L}$. This may sound a little odd, but it establishes the connection to a universal conventional mathematical terminology that we are now going to discuss.

Vertices

Definition 2.3.2
Let $\mathcal{L}$ be a map. A point in $\mathbb{R}^2$ is a *vertex* of $\mathcal{L}$ if it is the end point of an edge of $\mathcal{L}$.

The abstract notion of "vertex" now describes what we have, up to now, called either a "multinational corner" or a "border-stone." If $\mathcal{L}$ is a map, then, generally, its corresponding *set of vertices,* meaning the set of all vertices of $\mathcal{L}$, will be denoted by $E_{\mathcal{L}}$. The pair $G = (E_{\mathcal{L}}, \mathcal{L})$ then forms a finite *plane graph* in the sense of Harary [HARARY 1969, page 103].[5] In order to simplify matters, in what follows we will borrow from the customary presentation of the Four-Color Problem. A *graph* will always mean a finite plane graph, in most instances a graph without *isolated vertices* (that is, vertices that are not end points of edges [HARARY 1969, page 15]). In general, graph-theoretical terminology will be used. Therefore, the terms "vertex" and "edge" instead of "border-stone" and "arc" will be used. In keeping with this agreement, "map" and "graph" will also denote the same concept. From a map $\mathcal{L}$, one obtains the graph $(E_{\mathcal{L}}, \mathcal{L})$. From a graph $G = (E, \mathcal{L})$, one can retrieve the map $\mathcal{L}$ by eliminating the set of vertices. From now on, it is in this sense that we will use the terms "map" and "graph" almost synonymously.

Countries

At this point, we still don't even know what "countries" are! If $\mathcal{L}$ is a map, then we call all points that belong to an edge of $\mathcal{L}$ *neutral points*

---

[5]The definition given by Harary excludes the occurrence of multiple edges and loops in a graph. Such graphs are sometimes said to be simple. See also [WAGNER-BODENDIEK 1989, page 21].

of $\mathcal{L}$. The set of all neutral points, that is, the *neutrality set* of $\mathcal{L}$, will be denoted by $N_{\mathcal{L}}$, or, in the case where $\mathcal{L}$ is a fixed map, simply N. Because arcs are compact (Lemma 2.2.2) and since a union of finitely many compact sets is also compact, the neutrality set of a map is always compact. We also observe that every subset of edges of a map itself forms yet another map. Therefore, for every subset $\mathcal{L}'$ of a map $\mathcal{L}$, the vertex set $E_{\mathcal{L}'}$ and the neutrality set $N_{\mathcal{L}'}$ are well-defined.

Definition 2.3.3
Let $\mathcal{L}$ be a map. A *country* of $\mathcal{L}$ is a path component of the complement of the neutrality set of $\mathcal{L}$, that is, of $\mathbb{R}^2 \setminus N_{\mathcal{L}}$.

First we establish the following:

Theorem 2.3.4
Let $\mathcal{L}$ be a map and $\mathbf{x}$ a point in the plane that is not a neutral point of $\mathcal{L}$. All points that can be joined to $\mathbf{x}$ by an arc having no neutral points, together with $\mathbf{x}$ itself, form a country. ∎

Proof Let L^* be the set of all points in the plane that can be joined to $\mathbf{x}$ by an arc having no neutral points of $\mathcal{L}$. We must show that the set $L = L^* \cup \{\mathbf{x}\}$ is a path component of $\mathbb{R}^2 \setminus N_{\mathcal{L}}$. Because $\mathbf{x} \in L$, $L \neq \emptyset$. To prove path connectedness, it suffices to take two distinct points $\mathbf{y}_1, \mathbf{y}_2 \in L^*$. There exist arcs B_1 and B_2 without neutral points that join $\mathbf{x}$ to $\mathbf{y}_1$ and $\mathbf{y}_2$, respectively. We note that all points of B_1 and B_2 also belong to L. However, there exists an arc $B \subseteq B_1 \cup B_2$ that joins $\mathbf{y}_1$ to $\mathbf{y}_2$ (Proposition 2.2.6). Since B_1 and B_2 lie in L, B also lies entirely in L.

Now let two points $\mathbf{y} \in L$ and $\mathbf{z} \in (\mathbb{R}^2 \setminus N_{\mathcal{L}}) \setminus L$ be given. Moreover, let B be an arc that joins $\mathbf{y}$ to $\mathbf{z}$. We must show that B does not lie completely in $\mathbb{R}^2 \setminus N_{\mathcal{L}}$, meaning that B has a nonempty intersection with $N_{\mathcal{L}}$. For that, we choose an arc B' having no neutral points that joins $\mathbf{x}$ to $\mathbf{y}$. Then we can find an arc $B'' \subset B' \cup B$ that joins $\mathbf{x}$ and $\mathbf{z}$ (again Proposition 2.2.6). Since $\mathbf{z} \notin L$, B'' must intersect the set $N_{\mathcal{L}}$ nontrivially. As $B' \subset L$, such an intersection point must, however, belong to B. □

From this result, a few of the properties of maps, properties that were certainly anticipated, emerge.

Corollary 2.3.5

Let $\mathcal{L}$ be a map.
(a) *Every point of the plane that is not a neutral point belongs to exactly one country.*
(b) *A country is an open subset of the plane.*
(c) *There exists exactly one unbounded country.*   ∎

Proof   (a) Let $\mathbf{x} \in \mathbb{R}^2$ be a point that is not a neutral point. Denote by L the country obviously containing $\mathbf{x}$ that was described in the previous theorem. We assume that the point $\mathbf{x}$ belongs to yet another country L_1. Because of the path connectedness of L_1, every point of L_1 (distinct from $\mathbf{x}$) can be joined to $\mathbf{x}$ by an arc having no neutral points. By the construction, every such point, consequently, belongs to L. This implies $L_1 \subseteq L$. Analogously, it also follows that $L \subseteq L_1$. Therefore, $L_1 = L$.

(b) We must show that there exists, for each point $\mathbf{x}$ of a country L, a disk centered at $\mathbf{x}$ that is entirely contained in L. From part (a) of this corollary, it follows that we can take L to be the country described in Theorem 2.3.4. The point $\mathbf{x}$ is certainly not a neutral point. Because the neutrality set is closed, its complement is open, and we can find a disk centered at $\mathbf{x}$ that lies entirely in the complement. For each point $\mathbf{y}$ of the disk distinct from $\mathbf{x}$, the line segment from $\mathbf{y}$ to $\mathbf{x}$ is completely contained in this disk. Consequently, an arc without neutral points joins $\mathbf{x}$ and $\mathbf{y}$. The stated disk, therefore, is contained entirely in L.

(c) Since $N_{\mathcal{L}}$ is bounded, there exists a circle K such that $I(K) \supset N_{\mathcal{L}}$. We choose a point $\mathbf{x} \in A(K)$ and denote by L the only country containing $\mathbf{x}$. Every other point in $A(K) \cup K$ can be joined to $\mathbf{x}$ by an arc that contains no neutral point and that is contained entirely in $A(K) \cup K$. Then $A(K) \cup K \subset L$, and hence L, are both unbounded. Every other country is (by part (a)) disjoint from L. Therefore, each is contained in $I(K)$, and thereby each is bounded.   □

Connectedness of Maps

There is yet another meaning to the notion of connectedness.

Definition 2.3.6

A map is said to be *connected* if every pair of vertices can be joined by an arc consisting of concatenated edges.

This is precisely the case if the neutrality set is path connected. Because every subset of edges of a map $\mathcal{L}$ itself forms a map, the property of "connectedness" is also defined for arbitrary subsets of edges of a map. A nonempty subset $\mathcal{L}'$ of edges of a map $\mathcal{L}$ is a *component* of $\mathcal{L}$ if it itself is a connected map with the property that no larger subset containing it is also connected. Every map is a union of its components.

The difference in the two meanings of connectedness should be noted. The connectedness of countries belonging to a map $\mathcal{L}$ is a matter of the components of the complement of the neutrality set of the map. The connectedness of maps, on the other hand, depends upon the components of the neutrality set itself. In other words, a country is a component of the complement of the neutrality set; a component of a map corresponds to a component of the neutrality set itself.

The connectedness of a map involves a particular kind of connectedness of its corresponding countries. A country of a connected map can enclose no extraterritorial enclaves. A connected open set U in $\mathbb{R}^2$ is *simply connected* if every embedding (that is, every continuous injective mapping) of the unit circle S^1 into U can be extended to a continuous mapping of the *unit disk*

$$B^2 = \{\, (x,y) \in \mathbb{R}^2 \,:\, x^2 + y^2 \leq 1 \,\}$$

into U.[6]

Theorem 2.3.7

The bounded countries of a connected map are simply connected. ∎

Proof Let $\mathcal{L}$ be a connected map, and let L be a bounded country of $\mathcal{L}$. Suppose that $h : S^1 \to L$ is an embedding. Then the image of h

---

[6]The characterization of simple connectedness given here is formally weaker than that in the conventional literature in which the extension of all continuous mappings $S^1 \to U$ is required. However, this definition is, for open subsets of the plane, essentially equivalent.

is a closed Jordan curve K lying entirely in L. Since $\mathcal{L}$ is connected, the edges of $\mathcal{L}$ lie either completely in $I(K)$ or completely in $A(K)$. As L is bounded, the latter must hold. We can assume that $h(\mathbf{x}) = \mathbf{x}$ for all $\mathbf{x} \in S^1$ (Schoenflies theorem 2.2.7) and that all edges of $\mathcal{L}$ lie in the region exterior to the unit circle. Thus we can define an extension $h' : B^2 \to U$ given by $\mathbf{x} \mapsto \mathbf{x}$. $\square$

Reduction to Polygonal Arc Maps

Definition 2.3.8
A map is said to be a *polygonal arc map* if its edges are polygonal arcs having no self-intersections.

The first step in sidestepping the house of horrors of general arcs is provided by the following deep theorem.

Theorem 2.3.9
Let $\mathcal{L}$ be a map and U an open subset of the plane such that $N_{\mathcal{L}} \subset U$. Then there exists a homeomorphism $h : \mathbb{R}^2 \longrightarrow \mathbb{R}^2$ satisfying the following conditions:
1. *h maps $\mathcal{L}$ to a polygonal arc map.*
2. *h fixes the points not in U.*   ∎

This is a part of Theorem 8 in [MOISE 1977], whose proof we shall also omit. The proof requires the Jordan curve theorem (Theorem 2.2.5) and the Schoenflies theorem (Theorem 2.2.7).

Since connectedness of maps is preserved by homeomorphisms, we can, from a given map, always transfer to a polygonal arc map without altering the countries in an essential way. In particular, the coloring problem remains the same. Later, we will sharpen this result even more. We will, in fact, prove that we can even restrict ourselves to maps that consist only of line segments (Wagner and Fáry theorem 4.2.11).

Polygonal arc maps have a particular local structure that is helpful in many situations.

Theorem 2.3.10

If $\mathcal{L}$ is a polygonal arc map, then for every point $\mathbf{x}$ in the plane there exists a disk U centered at $\mathbf{x}$ whose intersection with the neutrality set N consists of finitely many radii of U. ∎

Proof We consider a fixed point $\mathbf{x}$ in the plane. Let N' be the subset of N consisting of all line segments that belong to a polygonal arc of $\mathcal{L}$ but do not contain $\mathbf{x}$. The set N' is a union of finitely many line segments, hence finitely many closed sets. Therefore, N' itself is closed. We conclude from this that $\mathbb{R}^2 \setminus N'$ is open. Consequently, for the given point $\mathbf{x}$, that belongs to $\mathbb{R}^2 \setminus N'$, there exists a disk U centered at $\mathbf{x}$ that is disjoint from N'. There are several possible cases to consider:

1. If $\mathbf{x} \notin N$, then $N' = N$, and so $U \cap N = \emptyset$. This means that $U \cap N$ consists of 0 radii.

2. If $\mathbf{x}$ is an interior point of a line segment S that belongs to a polygonal arc in $\mathcal{L}$, then N is formed from N' by including the interior points of S. From this, we obtain the following:

$$U \cap N = U \cap (N' \cup S) = (U \cap N') \cup (U \cap S) = U \cap S.$$

This set forms a diameter of U and consists of exactly two radii of U.

3. If $\mathbf{x}$ is an end point of a line segment belonging to a polygonal arc of $\mathcal{L}$, then $\mathbf{x}$ can also be the end point of other such line segments—however, only of finitely many of them. Denote these by $S_1, \ldots, S_n$. Then N is formed from N' by including $\mathbf{x}$ and the interior points of the line segments $S_1, \ldots, S_n$. The end points of the line segments $S_1, \ldots, S_n$ that are distinct from $\mathbf{x}$ already belong to N'. It follows that

$$U \cap N = U \cap \left(N' \cup \bigcup_{i=1}^{n} S_i\right) = (U \cap N') \cup \left(U \cap \bigcup_{i=1}^{n} S_i\right) = \bigcup_{i=1}^{n} (U \cap S_i)$$

and that each of the sets $U \cap S_i$ for $i = 1, \ldots, n$ is a radius of U. □

These facts point the way to a new concept.

Definition 2.3.11
Let $\mathcal{L}$ be a polygonal arc map. A neighborhood D of a point $\mathbf{x} \in \mathbb{R}^2$ is an *elementary neighborhood* of $\mathbf{x}$ (*relative to* $\mathcal{L}$) if it has the following properties:
1. D is a closed disk centered at $\mathbf{x}$.
2. $D \cap N_{\mathcal{L}}$ consists of finitely many radii of D.
3. No vertices of $\mathcal{L}$ lie on the circumference of D.

Theorem 2.3.10 implies that every point of the plane has elementary neighborhoods (relative to a given map). An almost immediate consequence of this fact is the earlier-mentioned (page 48) notion of arcwise accessibility of points of an arc from within a set.

Corollary 2.3.12
Every neutral point of a map $\mathcal{L}$ is arcwise accessible from within $\mathbb{R}^2 \setminus N_{\mathcal{L}}$. ∎

Proof Let $\mathcal{L}$ be a polygonal arc map and $\mathbf{x}$ a neutral point of $\mathcal{L}$. We choose an elementary neighborhood D of $\mathbf{x}$ and a point $\mathbf{y}$ in $D \setminus N_{\mathcal{L}}$. The line segment from $\mathbf{x}$ to $\mathbf{y}$ is an arc that, except for $\mathbf{x}$, is contained entirely in $\mathbb{R}^2 \setminus N_{\mathcal{L}}$. □

The existence of elementary neighborhoods is still guaranteed in a somewhat more restricted context.

Lemma 2.3.13
Let $\mathcal{L}$ be a polygonal arc map. Then every neighborhood of a point contains an elementary neighborhood of this point. ∎

Proof Let $\mathbf{x}$ be a point of the plane and U a neighborhood of $\mathbf{x}$. The definition of neighborhood yields a closed disk D_1 centered at $\mathbf{x}$ that is contained completely in U. At the same time, we also have an elementary neighborhood D_2 of $\mathbf{x}$ (Theorem 2.3.10). The smaller of the two concentric disks $D = D_1 \cap D_2$ is then an elementary neighborhood contained entirely in U. □

This lemma has an already long-recognized application.

Theorem 2.3.14
The neutrality set of a map is a nowhere dense subset of the plane. ∎

Proof The property of a set being "nowhere dense" (see page 48) is invariant under homeomorphisms. Consequently, by the proof of

Corollary 2.3.12, it suffices to consider a polygonal arc map $\mathcal{L}$. Let V be an open subset of the plane. We are looking for an open set V' that is contained entirely in V and that does not intersect the neutrality set of $\mathcal{L}$. If $V \cap N = \emptyset$, then we can take $V' = V$. Otherwise, we can find a point $\mathbf{x} \in V \cap N$ and, using the lemma proved above, an elementary neighborhood D of $\mathbf{x}$ that is contained entirely in V. The set N subdivides the disk D into finitely many sectors (this includes the possibility of there being only one sector, which would then cover an angle of 360°). We can now choose an open disk V' that lies entirely in one of these sectors and that has the required property. $\qquad\square$

With this last result, it is finally clear that our definition truly yields border "lines" in the obvious sense—without any thickenings at all. Before we continue with the theory, we will illustrate the previous development by a number of examples.

2.4   Basic Examples

A few of the following examples also show that the given definition of a map is more general than one at first imagines.

Example 2.4.1
It is not out of the realm of possibility that a map has no edges at all. That means that $\mathcal{L} = \emptyset$. In this case, $N = \emptyset$ as well, and the entire plane forms the only country of the *empty map*. $\qquad\square$

Example 2.4.2
Let $\mathcal{L}$ be a map with exactly one edge B. Then $N_{\mathcal{L}} = B$, and $\mathbb{R}^2 \setminus B$ is the only country. This is a special case of the characterization of maps having exactly one country (Theorem 2.4.4)—a fact we will now derive from Theorem 2.3.9 using a reduction to polygonal arc maps. $\qquad\square$

A map always has *at least* one country, namely the unbounded one (part (c) of Corollary 2.3.5). In maps having many edges, there will exist *precisely one* country if no subset of the existing edges forms a closed Jordan curve. Before we can make this statement precise, we must still introduce a few definitions. A map $\mathcal{K}$ is a *circuit* if its neutrality set $N_\mathcal{K}$ is a closed Jordan curve. The *interior domain* (*exterior domain*) of a circuit is the interior domain (exterior domain) of its neutrality set. We write $I(\mathcal{K})$ instead of $I(N_\mathcal{K})$, and $A(\mathcal{K})$ instead of $A(N_\mathcal{K})$, for short. Since two distinct edges of a map have at most one vertex in common (Definition 2.3.1), a circuit has at least three vertices. A circuit having exactly three edges is called a *triangle*. Edges that are the edges of a circuit in a map $\mathcal{L}$ are called the *circuit edges* of $\mathcal{L}$. A map is said to be *circuit-free* if it contains no circuit. A *tree* is a nonempty connected circuit-free map. A circuit-free map is called a *forest*, since its components are trees. Trees have specially designated vertices, called final vertices. A vertex of a map $\mathcal{L}$ is called a *final vertex* of $\mathcal{L}$ if it is the vertex of only one edge in $\mathcal{L}$. A *final edge* is one that is incident with a final vertex.

Lemma 2.4.3

A tree has at least two final vertices.   ∎

Proof by induction on the number of edges: Let $\mathcal{L}$ be a tree. Because a tree is not empty, we have at least one edge B_0. If B_0 is the only edge in $\mathcal{L}$, then the two end points of B_0 are the desired final vertices. This is the first stage of the induction process. For the induction step, we take the nonempty map $\mathcal{L}' = \mathcal{L} \setminus \{B_0\}$. There are two cases to consider:

1. $\mathcal{L}'$ is a tree. By the induction hypothesis, $\mathcal{L}'$ has at least two final vertices. If both end points of B_0 had been vertices of $\mathcal{L}'$, then $\mathcal{L}$ would have contained a circuit. Therefore, one end point of B_0 is a final vertex of $\mathcal{L}$, and the same is true for at least one final vertex of $\mathcal{L}'$.

2. $\mathcal{L}'$ is a forest consisting of (at least) two trees. Each of them, by the induction hypothesis, has at least two final vertices. At least one final vertex of each tree is also a final vertex of $\mathcal{L}$.   □

A final vertex of a component of a map $\mathcal{L}$ is also a final vertex of $\mathcal{L}$ itself. Every nonempty circuit-free map has at least one tree as a component and therefore also has final vertices. We maintain that there can also be edges in a map that are neither circuit edges nor final edges. Such edges will be termed *bridges*. The exclusion of a bridge B from a map $\mathcal{L}$ increases the number of its components. In other words, the bridge B links two components of the map $\mathcal{L} \setminus \{B\}$. Following this observation, we will now characterize maps having exactly one country.

Theorem 2.4.4

A map has one and only one country if and only if it is circuit-free.   ∎

Proof   First of all, we consider a map $\mathcal{L}$ that contains a circuit $\mathcal{K}$. No point of $A(\mathcal{K})$ can be joined to a point in $I(\mathcal{K})$ without intersecting $N_\mathcal{K}$ nontrivially (part 3 of the Jordan curve theorem 2.2.5). Therefore, no point in $A(\mathcal{K})$ can be joined to a point in $I(\mathcal{K})$ by an arc that has no neutral points. This implies that the unbounded country of $\mathcal{L}$ lies entirely in $A(\mathcal{K})$. In $I(\mathcal{K})$, there also exist points that are not neutral points (Theorem 2.3.14). Each such point belongs to a (bounded) country (part (a) of Corollary 2.3.5). Therefore, there must exist at least one bounded country, and hence there must be at least two countries in total. The given condition in the theorem is therefore necessary.

To prove the converse, we start with a polygonal arc map. First we make the observation that the number of countries of a map depends only upon the neutrality set of the map and not upon its particular subdivision into edges. Therefore, we can immediately, without essentially altering the situation, go directly to a map that consists of the line segments in the given existing polygonal arcs. Then it will suffice to show that circuit-free maps whose edges are all line segments have only one country. This can be done by induction on the number n of the existing line segments.

The case $n = 0$ is clear (see Example 2.4.1). Suppose the claim is true for circuit-free maps having n line segments. Let $\mathcal{L}$ be a circuit-free map consisting of $n + 1$ line segments. We choose a final vertex $\mathbf{e}$ of $\mathcal{L}$ and denote by S the corresponding final edge in $\mathcal{L}$. We also denote by $\mathcal{L}'$ the map formed from $\mathcal{L}$ by excluding S. This means that $\mathcal{L}' = \mathcal{L} \setminus \{S\}$. Furthermore, we set $N = N_\mathcal{L}$ and $N' = N_{\mathcal{L}'}$. We must now show that any two points in the complement of N can be joined by an arc that does not intersect N. To do that, let $\mathbf{x}_1$, $\mathbf{x}_2 \in \mathbb{R}^2 \setminus N$ be given. By the induction hypothesis, we can find an arc B' joining $\mathbf{x}_1$ and $\mathbf{x}_2$ that lies entirely in $\mathbb{R}^2 \setminus N'$. If B' does not intersect the line segment S, then we are finished. Otherwise, if $B' \cap S \neq \emptyset$, we modify B' in a suitable way. Let $\mathbf{z}$ be the point in $B' \cap S$

that is furthest away from **e**. Let S' be the part of the line segment S that joins **z** to **e**. Then the line segment S' lies entirely in the open set $U = \mathbb{R}^2 \setminus (N' \cup \{\mathbf{x}_1, \mathbf{x}_2\})$, and we choose a frame R for S' in U. Since the arc B' intersects the line segment S', we must, if we move along B' from $\mathbf{x}_1$ to $\mathbf{x}_2$, touch the frame several times, even traverse it. Now let $\mathbf{y}_1$ denote the point of R that we encounter first, and let $\mathbf{y}_2$ be the point at which we finally leave R without reentering again. Then we divide B' into three subarcs: B_a from $\mathbf{x}_1$ to $\mathbf{y}_1$, B_m from $\mathbf{y}_1$ to $\mathbf{y}_2$, and B_e from $\mathbf{y}_2$ to $\mathbf{x}_2$. We choose a polygonal arc B_r joining $\mathbf{y}_1$ to $\mathbf{y}_2$ that lies entirely in R and that is disjoint from S. This is possible because an end point of S lies in $I(R)$, and therefore S has at most one point in common with R. The concatenation of B_a, B_r, and B_e then constitutes an arc B that joins $\mathbf{x}_1$ to $\mathbf{x}_2$ and that does not intersect N. □

An interim result from the previous proof merits special mention.

Lemma 2.4.5
If $\mathcal{L}$ is a map and $\mathcal{K} \subset \mathcal{L}$ is a circuit, then there exists a country of $\mathcal{L}$ that is contained entirely in $I(\mathcal{K})$. ■

Example 2.4.6
In this context, the Jordan curve theorem (Theorem 2.2.5) implies the following: *A map that is a circuit has exactly two countries: a bounded one and an unbounded one.* □

We now want to produce a map with three countries. For that, we proceed from the geometric setting of Theorem 2.2.11.

Example 2.4.7
Let two points **x**, **y** be given. Let B_1, B_2, and B_3 denote three arcs joining these two points that pairwise have no interior points in common. Choose the indices such that $\overset{\circ}{B_2} \subset I(B_1 \cup B_3)$ (Theorem 2.2.11). Because these three arcs have the same end points, they do not form a map. We can construct one, however, by first placing suitable border-stones. We pick an interior point of each of B_1 and B_3. Thus we obtain four subarcs that together with B_2 yield a true map with four vertices and three countries (Theorem 2.2.11 and the proof of Corollary 2.2.12). □

Example 2.4.8

We choose three points on the unit circle. The map $\mathcal{L}$ consists of the three resulting arcs on the circle and the line segments from the center to the three chosen points.

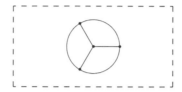

In this way, we obtain four countries that pairwise border on each other. Again, we have a map that cannot be admissibly colored with fewer than four colors. Later, we will see (Weiske's theorem 4.5.1) that a map can never have five countries that pairwise have common borderlines. □

Example 2.4.9

We pick n points on the unit circle where $n > 3$. The map $\mathcal{L}$ contains the n arcs on the circle and the line segments from the center to the n points in question. This results in $(n+1)$ countries, and the center is an n-nation vertex.

This example illustrates an important aspect in the treatment of the Four-Color Theorem, namely, that for an admissible coloring, different colors are required only for countries with a common "borderline." If one were to demand different colors for countries with only a "border point" in common, for such a map one would already need n colors for the bounded countries and one additional color for the unbounded country, which would be demarcated from the others by an arc of a circle. In this case, $n + 1 > 4$ colors in total would be required.

If n is even, then one can color the bounded countries of this map alternately with two colors. Since the unbounded country needs only one additional color, in this case, there would exist an admissible coloring with only three colors. If n is odd, however, an admissible coloring would necessitate four colors—as is the case in the previous example. □

2.5 National Borders

Starting only with our intuition, we have abstractly defined maps and countries. At this point, we will show that the properties of borders really emerge as we have always imagined them to be. At the end of this section, we will give a rigorous mathematical definition of the, up to now, intuitive notion of "common borderline." To do that, we shall assume (with no loss of generality because of Theorem 2.3.9) that all maps that we will consider in the proofs of this section and later in this chapter consist only of polygonal arcs.

Definition 2.5.1
Let $\mathcal{L}$ be a map and L a country of $\mathcal{L}$. A neutral point of $\mathcal{L}$ is called a border point of L if it is arcwise accessible from within L.

The path connectedness of countries (Definition 2.3.3) and the transitivity of path connectedness by means of arcs (Proposition 2.2.6) result in a variation of Corollaries 2.2.8 and 2.2.9.

Lemma 2.5.2
Let $\mathcal{L}$ be a map and L a country of $\mathcal{L}$.
(a) *A border point of L can be joined to every point of L by an arc whose interior points lie completely within L.*
(b) *Two border points of L can be joined by an arc all of whose interior points lie completely in L.* ∎

The countries of a map form disjoint open sets of the plane (part b) of Corollary 2.3.5). They contain, therefore, no boundary points. Moreover, no point in a country is a boundary point of another country. This means that the boundaries of every country are subsets of the neutrality set. What is more, the following is true:

Lemma 2.5.3
If $\mathcal{L}$ is a map and L a country of $\mathcal{L}$, then the border points of L are precisely the boundary points of L (in the topological sense). ∎

Proof Let $\mathbf{x}$ be a border point of L. Then we have an arc B one of whose end points is $\mathbf{x}$ and all of whose other points belong to L. Every disk centered at $\mathbf{x}$ contains $\mathbf{x}$ and points of $B \cap L$. Therefore, $\mathbf{x}$ is a boundary point of L.

Conversely, let $\mathbf{x}$ be a boundary point of L. Then $\mathbf{x}$ is, in any event, a neutral point of $\mathcal{L}$. We take an elementary neighborhood D of $\mathbf{x}$ that is subdivided by $N_{\mathcal{L}}$ into finitely many sectors, possibly two semicircles. Because $\mathbf{x}$ is a boundary point of L, one of these sectors must be contained in L and hence must be entirely contained in L up to the bounding radii. We now join one of the interior points of this sector to the center $\mathbf{x}$ (of the circle) with a line segment. This line segment is the arc needed to prove the arcwise accessibility of $\mathbf{x}$ from within L. □

On the basis of this lemma, it is unnecessary to introduce a separate definition for the set of border points of a country L. It is merely a question of the boundary of L, denoted by $B(L)$.

Remark: When we try to describe the boundaries of the countries in the examples from the previous section, then we stumble upon a few peculiarities. Caution is therefore recommended when one is using one's intuition.

- The boundary of the single country in the map having no edges (Example 2.4.1) is empty. However, every country of a map with edges has a nonempty boundary. One observes this in the following way:
  If one moves along the line segment connecting a point $\mathbf{x}$ of a country L to a certainly available neutral point $\mathbf{z}$, then one reaches a first point $\mathbf{y}$ that does not belong to L. It is possible that $\mathbf{y} = \mathbf{z}$, but not necessarily. In any case, $\mathbf{y}$ is a border point of L.

- A border point need not always be the border point of several distinct countries. In fact, this cannot be the case if a map having edges has only a single country (Example 2.4.2). ◇

The following assertion is important and intuitively clear, but it is able to be proved only with a certain amount of effort.

Theorem 2.5.4
Let $\mathcal{L}$ be a map. Then the boundary of each country of $\mathcal{L}$ is a union of edges of $\mathcal{L}$. ∎

The proof, which can be omitted for the moment, requires a preparatory result.

Lemma 2.5.5

Let $\mathcal{L}$ be a map, L a country of $\mathcal{L}$, and B an edge of $\mathcal{L}$. If an interior point of B is a border point of L, then the entire edge B belongs to the boundary of L. ■

Proof We must show that $B \subset B(L)$. By the assumption, we have a point $\mathbf{x} \in \overset{\circ}{B} \cap B(L)$. Assume that there exists a point $\mathbf{z} \in B \backslash B(L)$. Since the boundary $B(L)$ is a closed subset of the plane, we can find, as we stroll from $\mathbf{x}$ to $\mathbf{z}$ along B, a last point $\mathbf{y} \in \overset{\circ}{B} \cap B(L)$. We now consider an elementary neighborhood of $\mathbf{y}$. It will be subdivided by B into two sectors of which one will lie entirely in L. The two bordering radii of these sectors lie in $\overset{\circ}{B} \cap B(L)$. Therefore, since we must traverse one of these radii in our stroll towards $\mathbf{z}$, we do not exit the boundary of L at the point $\mathbf{y}$. □

Now the missing proof of the previous theorem follows easily.

Proof of Theorem 2.5.4 We must show that each border point of a country L belongs to an edge of $\mathcal{L}$ that is contained entirely in $B(L)$. The previous lemma takes care of the problem for border points of L that are interior points of edges. Now let the border point $\mathbf{x}$ of L be a vertex of $\mathcal{L}$ and therefore the end point of one or more edges of $\mathcal{L}$ (Definition 2.3.2). Once again, we consider an elementary neighborhood D of $\mathbf{x}$. Because $\mathbf{x}$ is a boundary point of L, one of the corresponding sectors of D must also lie in L. The bounding radii of such a sector are then contained in edges of $\mathcal{L}$ that belong to $B(L)$. The point $\mathbf{x}$ is a common end point of these edges and hence a point of those edges that lie entirely in $B(L)$. □

We are now able to introduce the following notion.

Definition 2.5.6

Let $\mathcal{L}$ be a map and L a country of $\mathcal{L}$.
(a) An edge that lies entirely in the boundary of L is called a *borderline* of L.
(b) The set of all borderlines of L, denoted by $\mathcal{G}_L$, is called the *border* of L.
(c) A set of edges of $\mathcal{L}$ is called a *national border* if it is the border of a country.

The border of a country is itself a map whose neutrality set is precisely the boundary of the country. We take note that the previously introduced notion of connectedness of maps has a special significance for national borders. One can distinguish between countries with connected borders and ones with borders that are not connected.

In what follows, we will also require a kind of converse to Theorem 2.5.4. From arcwise accessibility (Corollary 2.3.12), it follows that every neutral point of a map is the border point of a country. Over and above that, the following is true:

Theorem 2.5.7
An edge of a map belongs to at least one but at most two national borders.　■

Proof　Let B be an edge of a map $\mathcal{L}$, and let $\mathbf{x}$ be an interior point of B. It suffices to show that $\mathbf{x}$ is the border point of at least one, but at most two, countries (Lemma 2.5.5). We again consider an elementary neighborhood of $\mathbf{x}$. Since $\mathbf{x}$ is the interior point of an edge, the neighborhood will be subdivided into exactly two sectors by the neutrality set. These sectors can either belong to the same or to two different countries.　□

Later (Theorem 2.6.8), we will even characterize the edges of a map as to whether they belong to one or to two national borders. At this point, however, we state the following rather obvious assertion.

Corollary 2.5.8
A map has only finitely many countries.　■

We can improve this result by establishing that the number of countries of a map can be at most twice the number of edges (in the event that edges exist at all). That is, however, a very large upper bound. The Euler–Cauchy formula (Theorem 4.3.3) establishes a more exact estimate. This will be discussed in Chapter 4.

In terms of the structure of national borders, we can make two more precise statements.

Theorem 2.5.9
For maps having at least two countries, every national border contains at least one circuit.　■

Proof Let $\mathcal{L}$ be a map and L a country of $\mathcal{L}$ whose border $\mathcal{G}_L$ is circuit-free. Then a fixed point $\mathbf{x} \in L$ can be joined to each point in $\mathbb{R}^2 \setminus B(L)$ by an arc that does not intersect $B(L)$ (Theorem 2.4.4). Hence, $L = \mathbb{R}^2 \setminus B(L)$, implying that $\mathcal{L}$ has only one country, in contradiction to the assumption. $\square$

Theorem 2.5.10
A map that is not connected has a country whose border is not connected. ∎

Proof Let $\mathcal{L}$ be a map whose neutrality set N is not connected. We choose two points $\mathbf{x}_1$, $\mathbf{x}_2$ that lie in different components N_1, respectively N_2, of $\mathcal{L}$. Traveling along the line segment $[\mathbf{x}_1, \mathbf{x}_2]$ from $\mathbf{x}_1$ towards $\mathbf{x}_2$, we reach a last point belonging to N_1 and, after that, a first point that lies in a component of N different from N_1. The points in between lie in a country whose border is not connected. $\square$

2.6 Common Borderlines

Definition 2.6.1
Let $\mathcal{L}$ be a map. An edge is a common borderline of two countries of $\mathcal{L}$ if it belongs to the borders of both countries.

In general, there are many common borderlines.

Lemma 2.6.2
Every circuit edge of a map is a common borderline of two countries. ∎

Proof Let B be a circuit edge of a map $\mathcal{L}$. We pick a circuit $\mathcal{K}$ such that $B \in \mathcal{K} \subset \mathcal{L}$, a point $\mathbf{x} \in \overset{\circ}{B}$, and an elementary neighborhood D of $\mathbf{x}$ that will be subdivided into two sectors, in general semicircles, by B. These sectors (without the bounding radii) $D^i = D \cap I(\mathcal{K})$ and $D^a = D \cap A(\mathcal{K})$ are path connected and are disjoint from the set $N_{\mathcal{L}}$. Hence, each of them lies entirely in a country L^i, respectively L^a, of $\mathcal{L}$. Each arc that joins a point of D^i with a point of D^a must intersect $N_{\mathcal{K}}$, and therefore $N_{\mathcal{L}}$, nontrivially. This implies that the countries L^i and L^a are distinct from one another and have B as a common borderline. $\square$

Two different countries can have several common borderlines. We now describe a situation in which this is truly the case.

Lemma 2.6.3
Let **x** *be a vertex that is the end point of exactly two edges in the map* $\mathcal{L}$. *Let one of these two edges be a circuit edge. Then both edges are common borderlines of the same two countries.* ∎

Proof   We denote the two edges that are incident to the vertex **x** by B and B'. From what is given, we assume that B is a circuit edge. We choose a circuit $\mathcal{K} \subset \mathcal{L}$ that contains B. Because B' is the only edge in $\mathcal{L}$ besides B that contains **x**, B' must also belong to $\mathcal{K}$. This implies that B' is also a circuit edge. Now we choose an elementary neighborhood D of **x** that will be subdivided into two sectors $D^i = D \cap I(\mathcal{K})$ and $D^a = D \cap A(\mathcal{K})$ by the edges B and B', in a way similar to that described in the previous proof. Each of these sectors lies entirely in a single distinct country, and both edges B and B' are common borderlines of these two countries. □

The mutual positioning of the countries and the circuits in both of the above proofs will be described in the following definition.

Definition 2.6.4
Let $\mathcal{L}$ be a map. The circuit $\mathcal{K} \subset \mathcal{L}$ *separates* two countries of $\mathcal{L}$ if one of them lies in its interior domain and the other in its exterior domain.

The proof of the following fact is included in the proof of Lemma 2.6.2.

Lemma 2.6.5
If $\mathcal{K}$ *is a circuit of a map that contains a given circuit edge* B, *then it separates the countries whose borders contain* B. ∎

In fact, separating circuits exist in much more general situations.

Theorem 2.6.6
For two distinct countries of a map $\mathcal{L}$, *there always exists a separating circuit in* $\mathcal{L}$. ∎

Proof   Let L_1 and L_2 be two distinct countries of the map $\mathcal{L}$. We pick points $\mathbf{x}_1 \in L_1$ and $\mathbf{x}_2 \in L_2$. Every arc joining $\mathbf{x}_1$ to $\mathbf{x}_2$ must intersect the boundaries of both countries. Let us now remove an edge that is not a common borderline. We thus obtain a map $\mathcal{L}'$ for which the following conditions hold:

- The countries L_i, $i = 1, 2$, of $\mathcal{L}$ lie in countries L_i' of $\mathcal{L}'$.

- For at least one $i \in \{1, 2\}$, $L_i' = L_i$ and $B(L_i') = B(L_i)$.

Without loss of generality, we can assume that the latter condition holds for $i = 1$. Because every arc joining $\mathbf{x}_1$ to $\mathbf{x}_2$ has a nonempty intersection with the set $B(L_1) = B(L_1')$, the points $\mathbf{x}_1$ and $\mathbf{x}_2$ lie in different countries of $\mathcal{L}'$, implying that $L_1' \neq L_2'$. If there exists a circuit $\mathcal{K} \subset \mathcal{L}'$ of the type that one of the countries L_i' lies in $I(\mathcal{K})$ and the other in $A(\mathcal{K})$, then the circuit has this same property relative to the countries L^i. It suffices, therefore, to consider the countries L_i' of the map $\mathcal{L}'$. Now we can again remove an edge that is not a common borderline of the countries L_i'. In fact, we can continue doing this until, after finitely many steps, we obtain a map $\mathcal{L}^*$ that consists only of the common borderlines of the two countries L_i^*. However, we must be careful. In the first step, we are not permitted to remove at once all the edges that are not common borderlines of the countries L_i. Therefore, an edge can become a common borderline at a later stage even when it is not one at the outset.

Because every country of a nonempty map also has a nonempty border, it suffices from now on to prove the claim under the following additional assumptions: that there exist only two countries (Theorem 2.5.7) and that

$$\mathcal{L} = \mathcal{G}_{L_1} = \mathcal{G}_{L_2}.$$

Since the countries L_1 and L_2 are distinct from one another, $\mathcal{L}$ contains a circuit $\mathcal{K}$ (Theorem 2.5.9). In addition, because there must be one country of $\mathcal{L}$ lying in $I(\mathcal{K})$ and another in $A(\mathcal{K})$, there remains no other choice for the two single given countries except that one of them lies entirely in $I(\mathcal{K})$ and the other entirely in $A(\mathcal{K})$. $\qquad\square$

An important consequence of this result is:

Corollary 2.6.7
A common borderline of two countries of a map is always a circuit-edge. ∎

Proof Let $\mathcal{L}$ be a map. We consider a common borderline B of distinct countries L_1 and L_2 of $\mathcal{L}$, and we choose a separating circuit $\mathcal{K} \subset \mathcal{L}$. Without loss of generality, we can assume $L_1 \subset I(\mathcal{K})$ and $L_2 \subset A(\mathcal{K})$. If $B \not\subseteq \mathcal{K}$, then either $\overset{\circ}{B} \subset I(\mathcal{K})$ or $\overset{\circ}{B} \subset A(\mathcal{K})$. In the first instance, B can, however, not be a borderline of L_2. In the second case, it cannot be a borderline of L_1 (Lemma 2.5.2). Therefore, B is certainly not a common borderline. □

Now we finally have the long-awaited fine-tuning of Theorem 2.5.7, namely, the classification of edges according to the number of countries to whose borders they belong. From an intuitive point of view, it is straightforward.

Theorem 2.6.8
An edge of a map belongs to exactly two national borders if and only if it is a circuit edge. Bridges and final edges, respectively, belong to only one national border. ■

Proof The result comes directly from Lemma 2.6.2 and Corollary 2.6.7. □

An analysis of the proof of Corollary 2.6.7 also shows the following:

Lemma 2.6.9
Let two distinct countries of a map $\mathcal{L}$, together with a separating circuit $\mathcal{K}$, be given. Then the common borderlines of both countries belong to $\mathcal{K}$. ■

Definition 2.6.10
Two countries of a map that have a common borderline are said to be *neighboring countries*, or simply *neighbors*.

In view of the peculiarities pointed out at the beginning of this section, we observe that:

Lemma 2.6.11
There exists a neighboring country for each country of a map with at least two countries. ■

Proof Let L be a country of a map $\mathcal{L}$ having at least two countries. We are looking for a country that is a neighbor of L. Let $\mathbf{x}$ be a point of L and $\mathbf{z}$ a point of another country. We construct an arc B that joins $\mathbf{x}$ to $\mathbf{z}$ but that contains no vertex of $\mathcal{L}$. Such an arc must contain border points of L. The last of these, which we encounter by traveling along B starting from the point $\mathbf{x}$, must be an interior point of an edge belonging to the border of L and to one of the countries that are neighbors of L.

For the construction of such an arc B, we pick a positive real number r that is smaller than one-half the minimal distance between any two vertices of L and also smaller than all of the distances from the points $\mathbf{x}$ and $\mathbf{z}$ to each of the vertices. Then we begin with the line segment B_0 from $\mathbf{x}$ to $\mathbf{z}$. If B_0 ever passes through a vertex $\mathbf{y}$ of $\mathcal{L}$, we replace the diameter of the circle of radius r at $\mathbf{y}$ (which lies along B_0) by one of the two semicircular arcs joining the end points of this diameter. This is the desired arc B. □

2.7 The Extension of Maps

The following deliberations deal with the question of how countries and their borders change if one extends a given map by an edge. By induction, one can determine the behavior of maps that have been extended by a number of edges. In addition, the alterations to a map resulting from the elimination of edges can be inferred.

Let $\mathcal{L}$ be a map. Suppose that B is an arc with the property that the inclusion of B into $\mathcal{L}$ gives rise to a map $\mathcal{L}'$. This condition means that B has at most one or both of its end points in common with the neutrality set $N_{\mathcal{L}}$ and that such a common point must be a vertex of $\mathcal{L}$. As a result, there is exactly one country L of $\mathcal{L}$ such that $\overset{\circ}{B} \subset L$. The countries of the map $\mathcal{L}$ that are distinct from L are also countries of the map $\mathcal{L}'$. Moreover, in the transition from $\mathcal{L}$ to $\mathcal{L}'$, their borders do not alter. The country L, however, will change. There are three cases to consider:

Case 1: $B \cap N_{\mathcal{L}} = \emptyset$. ∎

Then B lies entirely in L and is, consequently, a final edge of $\mathcal{L}'$. If the open set $L' = L \setminus B$ is not path connected, then B must be a common borderline of two distinct countries of $\mathcal{L}'$. That cannot happen, however, because B is not a circuit edge of $\mathcal{L}'$ (Corollary 2.6.7). Hence, L' is a country of $\mathcal{L}'$ (Definition 2.3.3), meaning that the country L is reduced to the country L' having the border $\mathcal{G}_{L'} = \mathcal{G}_L \cup \{B\}$.

Case 2: $B \cap N_{\mathcal{L}} = \{\mathbf{x}\}$, where $\mathbf{x}$ is not only an end point of B but also a vertex of $\mathcal{L}$.   ■

Then the other end point of B lies in L, implying that B is a final edge of $\mathcal{L}'$. As B is not a circuit edge of $\mathcal{L}'$, the country L is again reduced to the country $L' = L \setminus B$ that has the border $\mathcal{G}_{L'} = \mathcal{G}_L \cup \{B\}$.

Case 3: $B \cap N_{\mathcal{L}} = \{\mathbf{x}, \mathbf{y}\}$, where $\mathbf{x}$, $\mathbf{y}$, the end points of B, are also vertices of $\mathcal{L}$.   ■

Now there are two possibilities. B can be a bridge or a circuit edge in $\mathcal{L}'$.

1. If B is a bridge in $\mathcal{L}'$, then the same thing happens as did before. L will be reduced to $L' = L \setminus B$ with $\mathcal{G}_{L'} = \mathcal{G}_L \cup \{B\}$.

2. If B is a circuit edge in $\mathcal{L}'$, then we can find a circuit $\mathcal{K}$ in $\mathcal{L}'$ such that $B \in \mathcal{K}$. The country L will be subdivided into the countries $L^i = L \cap I(\mathcal{K})$ and $L^a = L \cap A(\mathcal{K})$. The border of L^i consists of B, the borderlines of L whose interior points belong to $I(\mathcal{K})$, and possibly a few edges in $\mathcal{K} \cap \mathcal{G}_L$. The border of L^a consists of B, the borderlines of L whose interior points belong to $A(\mathcal{K})$, and the remaining edges in $\mathcal{K} \cap \mathcal{G}_L$.

In the latter case, we must still prove that L^i and L^a are countries of $\mathcal{L}'$ (see Definition 2.3.3). Certainly, only the path connectedness must be checked, since the other conditions in the definition are clearly fulfilled. We can assume that $I(\mathcal{K})$ is the upper hemisphere of the unit disk, that is,

$$I(\mathcal{K}) = \{(s, t) \in B^2 \setminus S^1 : t > 0\}.$$

We can also assume that B is the diameter of the unit circle lying on the boundary of this surface (Schoenflies theorem 2.2.7). Then we consider two points $\mathbf{u}, \mathbf{v} \in L^i$. Because these points lie in L, we can find an arc B' joining them that lies entirely in L. If B' lies completely within L^i, then we are finished. Otherwise, there exists, along the path from **u** to **v**, a first point $\mathbf{u}' \in B' \cap \overset{\circ}{B}$ and a last point $\mathbf{v}' \in B' \cap \overset{\circ}{B}$. The line segment $[\mathbf{u}', \mathbf{v}']$ lies entirely in the open set $L \setminus \{\mathbf{u}, \mathbf{v}\}$, and there exists a frame R for it in that set (Lemma 2.2.13). Now we alter the arc B' to form another arc B'' in the following way. We travel from **u** along B' until we meet, in R, a point with a positive second coordinate. Then we continue along R, always through points with positive second coordinates, until we reach the last point that B' has in common with R. From there, we again continue along B' to **v**. Thus we have an arc B'' joining **u** to **v** that lies entirely in L^i. This proves the path connectedness of L^i. A similar proof illustrates the path connectedness of L^a.   □

We close this section with an application of the preceding ideas. This application is significant mainly in the construction of "dual" maps (Section 4.4).

Proposition 2.7.1
Let $\mathcal{L}$ be a map and L a country of $\mathcal{L}$. Furthermore, let $\mathbf{x}$ be a point in L, and let finitely many border points $\mathbf{y}_1$, $\mathbf{y}_2$, ..., $\mathbf{y}_n$ of L be given. Then $\mathbf{x}$ can be joined to each $\mathbf{y}_i$, respectively, with an arc B_i that lies entirely in L except for the end point $\mathbf{y}_i$. These arcs may also be chosen such that pairwise they have no interior points in common.   ∎

Proof The subdivision of edges of a given map does not change the arrangement of countries in the plane. Therefore, without loss of generality, one can assume that the points $\mathbf{y}_i$ are vertices of $\mathcal{L}$. Because $\mathbf{y}_1$ is a border point of L, first of all, we can find an arc B_1 from **x** to $\mathbf{y}_1$ that except for the end point $\mathbf{y}_1$ lies entirely in L (part (a) of Lemma 2.5.2). The set $\mathcal{L}_1 = \mathcal{L} \cup \{B_1\}$ again forms a map. This map has the same countries as $\mathcal{L}$—except for L, which is reduced to the country $L_1 = L \setminus B_1$ (see Case 2 on page 82). The points **x** and $\mathbf{y}_2$ are border points of L_1 and can be joined by an arc B_2 that except for its end points lies entirely in L_2 (part (b) of Lemma 2.5.2). Next, we form the map $\mathcal{L}_2 = \mathcal{L}_1 \cup \{B_2\}$. The country L_1 will then either be reduced to the country $L_2 = L_1 \setminus B_2$ or be partitioned into two countries L_2', L_2'' having B_2 as a common borderline (see Case 3 on page 82). Now the points **x** and $\mathbf{y}_3$ are border points of one of the existing countries and can be joined together by a simple arc B_3 that except for its end points lies entirely in this country. This procedure can be continued until one has finally found the arc B_n. It is important only that **x** is a border point of *all* the countries lying inside L that are created by this process and that each point y_i is always a border point of *one* of the countries into which L is, after a certain number of steps, subdivided. □

3 CHAPTER

The Four-Color Theorem (Topological Version)

3.1 Formulation and Basic Approach

Generally, when one talks about four specific colors, then one frequently chooses the colors *blue, yellow, green*, and *red*.[1] For the most part, we will designate these four colors by the numbers 1, 2, 3, 4, and sometimes also by 0, 1, 2, 3. This has an additional advantage. One can then extend the discussion to more than four colors without any further difficulty. In fact, one is able to color with n colors, where n is any natural number ($n \in \mathbb{N}$). If $\mathcal{L}$ is a map, we denote by $\mathcal{M}_{\mathcal{L}}$ the set of all countries of $\mathcal{L}$. By "coloring" of a map, we intuitively mean that we have one color corresponding to each country of the map (see page 44).

---

[1] This choice is—in the English-speaking world, in any case—so standardized that it has given rise to a play on words. It relates to K. Appel and W. Haken, who solved the Four-Color Problem. Apples are red and green; so Appel is the "red–green" partner. Therefore, Haken must be the "blue–yellow" one.

Definition 3.1.1

Let $\mathcal{L}$ be a map and $n \in \mathbb{N}$. An *n-coloring* of $\mathcal{L}$ is a mapping $\varphi : \mathcal{M}_{\mathcal{L}} \to \{1, \ldots, n\}$. An *n*-coloring is *admissible* if neighboring countries always have distinct function values (i.e., "colors").

In the course of our deliberations, a "recoloring" will sometimes be required. One trivial instance of that will now be described.

Lemma 3.1.2

Let $\varphi : \mathcal{M}_{\mathcal{L}} \to \{1, \ldots, n\}$ be an admissible n-coloring of a map $\mathcal{L}$ and $\pi : \{1, \ldots, n\} \to \{1, \ldots, n\}$ a permutation (bijective mapping). Then the composition $\pi \circ \varphi$ is also an admissible n-coloring. ∎

We consider two colorings of a map to be *equivalent* if they differ only by a permutation of colors.

We are now able to formulate the Four-Color Theorem itself.

Theorem 3.1.3   (Four-Color Theorem)

For every map there exists an admissible 4-coloring. ∎

The basic approach to the exceedingly difficult proof of this theorem is really quite simple. It is an out-and-out standard variation of classical induction that involves the investigation of "minimal counterexamples"—sometimes referred to as "minimal criminals" or "smallest criminals." The following ideas form a basis for the approach. If there exist maps that cannot be colored with four colors, then there must be one such map having the fewest number f of countries. Because maps with at most four countries can evidently be colored with four colors, f must be > 4, and any map with fewer than f countries has a 4-coloring. We call a map with f countries that cannot be colored with four colors a *minimal counterexample* or a *minimal criminal*.[2] The entire thrust of the proof consists in showing that there cannot exist such a minimal criminal.

---

[2]From the work of [ROBERTSON et al. 1997], we learned that it would have been more clever to search for counterexamples in which the sum of the numbers of countries and edges is minimal. However, the advantage of that approach is not great enough to warrant adapting the remainder of this book to it.

3.2   First Steps Towards the Proof

As was mentioned earlier, the crux of the Four-Color Theorem is primarily of a combinatorial nature. A few of the properties satisfied by a minimal counterexample can, however, be derived in this topological setting. To whet the appetite, so to speak, we will derive these properties immediately. A reader who, on the first reading, has skipped over the topological foundations in the previous chapter can begin here if he/she familiarizes him/herself with the relevant notions by using the index.

At the moment, we will focus on a few maps that clearly have a "clean slate" and therefore cannot be minimal criminals. Thus we will be excluding them in any of our future hunts for criminals.

Proposition 3.2.1
In a minimal criminal there is no country that has fewer than four neighbors.   ∎

Proof Since a minimal criminal has at least five countries, each country has at least one neighbor (Lemma 2.6.11). If there were to exist a country L with at most three neighbors, then L could be merged with a neighboring country L' (by removing one of its common borderlines). Thus one obtains a map with one country fewer and that by assumption can be colored with four colors. Now put the deleted border back again and keep the color of the previously merged country for L'. Then one has in any case one color free for L.   □

In tackling the Four-Color Theorem, we could therefore limit ourselves to maps in which each country has at least four neighbors. In doing so, a country like San Marino, which lies entirely inside of another country, namely Italy, would have to be excluded. This condition would also exclude a country like Andorra, which has only France and Spain as neighbors. Another encompassing approach to the Four-Color Problem consists in establishing that if there exists a minimal criminal at all, it must have certain properties. In other words, it is sufficient to limit the search for criminals to maps that fulfill certain additional criteria. Naturally, we can limit ourselves to polygonal arc maps. We will now derive another such property.

As was the case in the preceding proposition, it involves some kind of numerical criterion, but for that it is convenient to have another graph-theoretical concept at our disposal.

Definition 3.2.2
Let **x** be a vertex of a map $\mathcal{L}$. The number of edges of $\mathcal{L}$ that have **x** as an end point is called the *degree* of **x** in $\mathcal{L}$ and will be denoted by $d_{\mathcal{L}}(\mathbf{x})$.

From an abstract point of view, we have simply defined a function $d_{\mathcal{L}} : E \to \mathbb{N}$ from the vertex set E into the set of all natural numbers. A final vertex has degree 1. All other vertices of a map (note that by our definition a map is a graph with no isolated points) have degree ≥ 2. Finally, we will use any one of the following expressions to indicate that a vertex is of degree d: *vertex of degree d, vertex with degree d, d-vertex*. Be careful: One cannot interchange the notions of *triangle* and *3-vertex*, *square* and *4-vertex*, *n-gon* and *n-vertex*!

We will now rule out a few more maps in our search for counterexamples.

Lemma 3.2.3
If there exists a minimal criminal with f countries, then there exists one with f countries that has no bridges or final edges.   ∎

Proof   In admissible colorings of a map, only the edges that are the common borderlines of two distinct countries play a roll, that is, only circuit edges (Corollary 2.6.7). By removing a bridge or a final edge from a map, it is possible to create final edges from other bridges. However, the nature of any of the remaining edges, whether it is a circuit edge or not, does not change. Consequently, by removing all available bridges and final edges from a minimal criminal, one obtains another minimal criminal having no bridges or final edges.   □

Next we show:

Lemma 3.2.4
A minimal criminal having no bridges or final edges is a connected map.   ∎

Proof Let $\mathcal{L}$ be a minimal criminal with no bridges or final edges. Assume that $\mathcal{L}$ is not connected. Then there exists a country L whose border $\mathcal{G}_L$ is not connected (Theorem 2.5.10). By a double application of the stereographic projection, we can assume that L is the unbounded country of $\mathcal{L}$. We now choose a circuit $\mathcal{K}_1 \subset \mathcal{G}_L$ (Theorem 2.5.9). L is then contained within its exterior domain $A(\mathcal{K}_1)$. From $\mathcal{L}$, we form the map $\mathcal{L}_1$ by deleting the components of $\mathcal{L}$ that lie entirely in $A(\mathcal{K}_1)$. We also define the map $\mathcal{L}_2 := \mathcal{L} \setminus \mathcal{L}_1$. Let L_1 and L_2 denote the unbounded countries of $\mathcal{L}_1$ and $\mathcal{L}_2$, respectively. We note that $\mathcal{L}_1$ and $\mathcal{L}_2$ contain only circuit edges.

Because $\mathcal{G}_L$ is disconnected but is composed of only circuit edges, we can find another circuit $\mathcal{K}_2 \subset \mathcal{L}_2$. Some countries of the map $\mathcal{L}$, distinct from L itself, still lie in $A(\mathcal{K}_1)$, possibly even in $I(\mathcal{K}_2)$. In the transition from $\mathcal{L}$ to $\mathcal{L}_1$, these are merged with L to form L_1. Consequently, $\mathcal{L}_1$ has fewer countries than $\mathcal{L}$. Thus we can find a 4-coloring φ_1 of $\mathcal{L}_1$ in which we can, by recoloring if necessary (Lemma 3.1.2), choose $\varphi_1(L_1) = 1$.

In $I(\mathcal{K}_1)$ as well there are countries distinct from L. These will, in the transition from $\mathcal{L}$ to $\mathcal{L}_2$, be amalgamated with L to form L_2. Therefore, $\mathcal{L}_2$ also has fewer countries than $\mathcal{L}$, and we can find a 4-coloring φ_2 of $\mathcal{L}_2$ such that $\varphi_2(L_2) = 1$.

Now, a country of $\mathcal{L}$ distinct from L itself is a bounded country of either $\mathcal{L}_1$ or $\mathcal{L}_2$. However, no bounded country of $\mathcal{L}_1$ (as a country of $\mathcal{L}$) is a neighbor of a bounded country of $\mathcal{L}_2$ (again, considered as a country of $\mathcal{L}$). Thus we can obtain a 4-coloring of $\mathcal{L}$ in which we color L with the color 1, any bounded country in $\mathcal{L}_1$ according to the coloring determined by φ_1, and any bounded country in $\mathcal{L}_2$ in conformance with φ_2. Hence, $\mathcal{L}$ is not a minimal criminal. $\qquad\square$

Both of the following technical properties will be needed for the next deep theorem.

Lemma 3.2.5
Let $\mathcal{L}$ be a minimal criminal. Then the following properties hold:
(a) *There exists no circuit $\mathcal{K} \subset \mathcal{L}$ with exactly three edges for which both the interior domain and the exterior domain of $N_\mathcal{K}$ contain more than one country of $\mathcal{L}$.*
(b) *Let $\mathbf{x}$ be a 2-vertex of $\mathcal{L}$ and let B_1, B_2 be the two edges of $\mathcal{L}$ that are incident with $\mathbf{x}$. Then the end points of B_1 and B_2 that are distinct*

from **x** *are not joined by an edge of* $\mathcal{L}$. *This means that the set*

$$\mathcal{L}' = (\mathcal{L} \setminus \{B_1, B_2\}) \cup \{B_1 \cup B_2\}$$

is again a map and is itself a minimal criminal.   ■

Proof of (a)   Let $\mathcal{L}$ be a minimal criminal and $\mathcal{K} = \{B_1, B_2, B_3\}$ a circuit in $\mathcal{L}$. For $j = 1, 2, 3$, L_j^i will denote the country lying in $I(\mathcal{K})$ to whose border the edge B_j belongs; L_j^a will be the country lying in $A(\mathcal{K})$ to whose border the edge B_j belongs (Lemmas 2.6.2 and 2.6.5). The claim implies that either the three countries L_j^i coincide or the three countries L_j^a coincide, but not both. To prove this by contradiction, we assume that there are not only two distinct L_j^i but also two distinct L_j^a.

As we did in the previous proof, we transfer to maps with fewer countries than the original map $\mathcal{L}$:

$$\mathcal{L}^i := \{B \in \mathcal{L} : \overset{\circ}{B} \not\subset A(\mathcal{K})\},$$

$$\mathcal{L}^a := \{B \in \mathcal{L} : \overset{\circ}{B} \not\subset I(\mathcal{K})\}.$$

Then $A(\mathcal{K})$ is a country of $\mathcal{L}^i$, and in the transition from $\mathcal{L}$ to $\mathcal{L}^i$, all the countries of $\mathcal{L}$ that lie in $A(\mathcal{K})$ will be merged with $A(\mathcal{K})$. Since $\mathcal{L}$ has at least two such countries lying in $A(\mathcal{K})$, $\mathcal{L}^i$ has fewer countries than $\mathcal{L}$, and we can find a 4-coloring φ^i of $\mathcal{L}^i$. Analogously, $I(\mathcal{K})$ is a country of $\mathcal{L}^a$. In the transition from $\mathcal{L}$ to $\mathcal{L}^a$, all countries of $\mathcal{L}$ that lie in $I(\mathcal{K})$ will be merged with $I(\mathcal{K})$. Since $\mathcal{L}$ also has at least two countries lying in $I(\mathcal{K})$, $\mathcal{L}^a$ similarly has fewer countries than $\mathcal{L}$, and we can find a 4-coloring φ^a of $\mathcal{L}^a$. We now show that we can, by recoloring, always guarantee that

$$\varphi^i(L_j^i) \neq \varphi^a(L_j^a)$$

for $j = 1, 2, 3$. To do that, we must distinguish between several cases:

1. At most two colors are required for the coloring of not only the L_j^i but also the L_j^a through the mapping φ^i, respectively φ^a. By recoloring (Lemma 3.1.2), we can then color the countries L_j^i with the colors 1, 2 and the countries L_j^a with the colors 3, 4.

2. Three colors are needed for both the L_j^i and the L_j^a through φ^i and φ^a, respectively. By recoloring, we can then arrange that $\varphi^i(L_j^i) = j$ and $\varphi^a(L_j^a) = j + 1$ for $j = 1, 2, 3$.

3. Three colors are required for the countries L_j^i and only two for the countries L_j^a. We can assume that $\varphi^i(L_j^i) = j$ for $j = 1, 2, 3$ and that

$$\varphi^a(L_1^a) = \varphi^a(L_2^a) \neq \varphi^a(L_3^a).$$

By recoloring, we can set $\varphi^a(L_1^a) = \varphi^a(L_2^a) = 3$ and $\varphi^a(L_3^a) = 4$. We proceed analogously in the case that three colors are needed for the L_j^a's and only two for the L_j^i's.

4. Three colors are needed for the L_j^i and only one for the L_j^a, $j = 1, 2, 3$. Then we can, by recoloring, arrange that $\varphi^i(L_j^i) = j$ for $j = 1, 2, 3$ and $\varphi^a(L_j^a) = 4$. We proceed in a similar fashion if the L_j^a's require three colors and the L_j^i's require only one.

In this way, the desired result is achieved.

Now, no country lying in $I(\mathcal{K})$ that is distinct from the countries L_j^i can be a neighbor of a country lying in $A(\mathcal{K})$ distinct from the L_j^a's. Because of this, we can obtain a 4-coloring of $\mathcal{L}$ in which the countries lying in $I(\mathcal{K})$ are colored in accordance with φ^i and the countries contained in $A(\mathcal{K})$ according to φ^a. Hence, $\mathcal{L}$ cannot be a minimal criminal.

Proof of (b) Let $\mathbf{x}$ be a vertex of $\mathcal{L}$ that is the end point of exactly two edges $B_1, B_2 \in \mathcal{L}$ whose other end points are joined by a third edge $B_3 \in \mathcal{L}$. The edges $B_j, j \in \{1, 2, 3\}$, then form a circuit $\mathcal{K} \subset \mathcal{L}$. Moreover, there exist countries $L^i \subset I(\mathcal{K})$, $L^a \subset A(\mathcal{K})$ having B_1, B_2 as common borderlines (Lemmas 2.6.2 and 2.6.3). As well, there are countries $L^{i'} \subset I(\mathcal{K})$, $L^{a'} \subset A(\mathcal{K})$ with B_3 as a common borderline (Lemmas 2.6.2 and 2.6.5).

Now, if no edges $B \in \mathcal{L}$ were to exist such that $\overset{\circ}{B} \subset I(\mathcal{K})$, then $L^i = L^{i'}$, and this country would have only two neighbors L^a and $L^{a'}$. This is not possible in maps that are minimal criminals (Proposition 3.2.1). In an analogous way, we can treat the case when there are no edges $B \in \mathcal{L}$ with the property that $\overset{\circ}{B} \subset A(\mathcal{K})$.

By part (a), it suffices to show that $I(\mathcal{K})$ will be partitioned by $\mathcal{L}$ into at least two countries. Analogously, this will also then be the case for $A(\mathcal{K})$. We already know that there is at least one edge B such that $\overset{\circ}{B} \subset I(\mathcal{K})$. Because all edges are circuit edges, we can also find a circuit $\mathcal{K}'$ that contains B. If $\overset{\circ}{B}' \subset I(\mathcal{K})$ for all $B' \in \mathcal{K}'$, first of all, it follows that the closed Jordan curve $N_{\mathcal{K}'}$ is contained in $I(\mathcal{K}) \cup N_{\mathcal{K}'}$. Secondly, $A(\mathcal{K})$ is contained in $A(\mathcal{K}')$. Thirdly, $I(\mathcal{K}) \supset I(\mathcal{K}')$, and, finally, $L^i \cup L^{i'} \subset A(\mathcal{K}')$. It could even happen that $L^i = L^{i'}$ throughout, but, in any case, $I(\mathcal{K}')$ encloses another country that is distinct from L^i and $L^{i'}$. (Theorem 2.2.11). Thus we have again at least two countries in $I(\mathcal{K})$. □

We are now in a position to obtain the following result.

Theorem 3.2.6
If there exists a minimal criminal $\mathcal{L}$ with f countries, then there exists one with f countries such that each vertex has at least degree 3. ■

Proof Let $\mathcal{L}$ be a minimal criminal that has no bridges and no final edges (Lemma 3.2.3). In the definition of a map (Definition 2.3.1), we have excluded isolated vertices, and hence 0-vertices. $\mathcal{L}$ has no 1-vertices because 1-vertices can only arise if the map has final edges. We can eliminate each 2-vertex by merging its adjacent edges (part (b) of Lemma 3.2.5). Note that if several such vertices exist, then we must go through this process several times over. The end result is a minimal criminal $\mathcal{L}'$ with the desired property. □

One more restriction on the maps under scrutiny comes from the following claim.

Lemma 3.2.7
Consider a map $\mathcal{L}$ that is a minimal criminal having no bridges. Suppose that $\mathcal{L}$ also satisfies the property that each of its vertices is of degree ≥ 3. Then two distinct countries of $\mathcal{L}$ have at most one common borderline. ■

Proof Let $\mathcal{L}$ be a minimal criminal that satisfies the given conditions. Suppose there exist two countries L_1, L_2 of $\mathcal{L}$ that have two distinct common borderlines B and B'. First, we choose a circuit $\mathcal{K} \subset \mathcal{L}$ that separates the two countries (Theorem 2.6.6). It will contain the edges B and B' (Lemma 2.6.9). Then we pick two points $\mathbf{x}$, $\mathbf{x}'$ that are interior to B, respectively B'. Finally, for $j = 1, 2$, we pick arcs B_j that join the points $\mathbf{x}$ and $\mathbf{x}'$ and that satisfy the property that $\overset{\circ}{B_j} \subset L_j$. Then $K_{12} = B_1 \cup B_2$ is a closed Jordan curve. Now we surmise that perhaps both the interior domain and the exterior domain of K_{12} contain entire countries of $\mathcal{L}$. These countries would certainly be distinct from L_1 and L_2.

It suffices to verify this for $I(K_{12})$. The result for $A(K_{12})$ follows analogously. Towards that end, we are looking for a circuit $\mathcal{K}'' \subset \mathcal{L}$ that is contained entirely in $I(K_{12})$ and that encompasses a country of the desired type (Lemma 2.4.5).

The closed Jordan curve $N_{\mathcal{K}}$ will be subdivided into two subarcs B_l and B_r by the points $\mathbf{x}$ and $\mathbf{x}'$. Each of these arcs will contain an end point of B and an end point of B'. Because one of the arcs B_j lies in $I(\mathcal{K})$ and the other in $A(\mathcal{K})$ (up to the end points), then one of the arcs B_l, B_r (again, except for the end points) lies in $I(K_{12})$, the other in $A(K_{12})$ (Corollary 2.2.10). Thus B and B' each have exactly one end point lying in $I(K_{12})$. Let $\mathbf{y}$ denote the end point of B lying in $I(K_{12})$, and let $\mathbf{y}'$ denote the end point of B' lying in $I(K_{12})$. By assumption, $d_{\mathcal{L}}(\mathbf{y}) \geq 3$. Hence, we can find an edge B'' distinct from B and B' having $\mathbf{y}$ as an end point. Again, by assumption, B'' is a circuit edge. We now claim that B'' belongs to a circuit $\mathcal{K}'' \subset \mathcal{L}$ all of whose edges lie in $I(K_{12})$.

In order to see this, first of all we choose an arbitrary circuit $\mathcal{K}' \subset \mathcal{L}$ containing B''. Starting from $\mathbf{y}$, we then travel along the simple curve $N_{\mathcal{K}'}$ beginning with the edge B''. Because $N_{\mathcal{K}'}$ is a closed Jordan curve, we must at some time or other, at the very latest at the end of our journey, reach a vertex of $\mathcal{K}$. Proceeding in the direction of travel, let $\mathbf{y}''$ be the first such vertex that we encounter. The points $\mathbf{x}$ and $\mathbf{x}'$ offer the only possibilities for one to exit the interior domain of K_{12} along the neutrality set of $\mathcal{L}$. To reach these points, we must first pass through the vertices $\mathbf{y}$ and $\mathbf{y}'$. In any case, up to the point $\mathbf{y}''$, we are still moving entirely within $I(K_{12})$. If $\mathbf{y}'' = \mathbf{y}$, then the curve is already closed, and we set $\mathcal{K}'' = \mathcal{K}'$. Otherwise, we modify $\mathcal{K}'$ by deleting the as yet untraveled edges and by replacing theM with the edges of $\mathcal{K}$ in $I(K_{12})$ that join $\mathbf{y}''$ to $\mathbf{y}$. In this way, we obtain the desired circuit $\mathcal{K}''$.

Thus the specified interim result has been achieved. Now we construct two new maps $\mathcal{L}^i$ and $\mathcal{L}^a$. $\mathcal{L}^i$ is produced from $\mathcal{L}$ by removing all edges lying entirely in $I(K_{12})$ and, in the case that $\mathbf{y} \neq \mathbf{y}'$, by adding, as edges, the subarcs of $\mathcal{K}$ that lie in $I(K_{12})$. In doing this, the countries lying in $A(K_{12})$ are not altered, but at least one of the countries L_1, L_2 will be enlarged. Every country of $\mathcal{L}$ lying entirely in $I(K_{12})$ will be merged with either L_1 or L_2. Because $\mathcal{L}$ is a minimal criminal and $\mathcal{L}^i$ contains at least one country less than $\mathcal{L}$, we can color $\mathcal{L}^i$ with four colors. Moreover, we can do it in such a way that for $j = 1, 2$, the (enlarged) country L_j is colored with the color j. In a similar fashion, we can create and color the map $\mathcal{L}^a$. The two colorings taken together yield an admissible 4-coloring of $\mathcal{L}$. This implies that $\mathcal{L}$ cannot be a minimal criminal, and we have the required contradiction.   □

The preceding deliberations about the Four-Color Theorem are now collected together into the following definition.

Definition 3.2.8
A map is said to be *regular* if it fulfills the following conditions:
1. It is not empty.
2. It is connected.
3. It contains no bridges and no final edges.
4. Any two distinct countries have at most one common borderline.

Now we can formulate the following theorem.

Theorem 3.2.9
If there exist minimal criminals at all, then there must be regular maps among them.[3]   ■

Proof Let $\mathcal{L}$ be a minimal criminal. Since it must have at least five countries, it is not empty. In particular, it has circuit edges. Furthermore, we can assume that it is without bridges and final edges (Lemma 3.2.3). In this instance, however, $\mathcal{L}$ is also connected (Lemma 3.2.4). The amalgamation of two circuit edges that abut at a 2-vertex is always possible (part (b) of Lemma 3.2.5). This process again yields a circuit edge. Therefore, we can assume that $\mathcal{L}$ contains only vertices of degree ≥ 3. In this case, however, two distinct countries have at most one common borderline (Lemma 3.2.7).   □

---

[3]If we had taken the alternative definition of a minimal criminal as described in footnote 2 on page 86, the statement of this theorem could have been simplified to the following: *A minimal criminal, if it exists, is a regular map.*

This theorem means that we can in our analyses limit our search for minimal criminals to maps that are regular. By doing this, we are making an agreement. From now on, when we mention a minimal criminal, we will always tacitly assume that we are dealing with a regular map.

We expressly point out, however, that nonregular maps may nonetheless crop up in the pursuit of minimal criminals. If this occurs, the maps in question would not be minimal counterexamples. We take note that the condition of regularity, in general, imposes a lower bound on the possible degrees of all vertices, a fact we found earlier about minimal criminals (Theorem 3.2.6).

Lemma 3.2.10
Every vertex of a regular map has at least degree 3. ■

Proof The absence of final edges in a regular map precludes the existence of 1-vertices. Since any two distinct countries have at most one common borderline and since all edges are circuit edges, there are also no 2-vertices (Lemma 2.6.3). □

In addition to this, we obtain the following far-reaching result.

Lemma 3.2.11
A regular map has at least four countries. ■

Example 2.4.8 illustrates that this lower bound on the number of countries in a regular map is really sharply defined.

Proof Let $\mathcal{L}$ be a regular map. As $\mathcal{L} \neq \emptyset$, it has at least one edge, and because of the exclusion of bridges and final edges, this edge must be a circuit edge. Therefore, we have at least two countries (Lemma 2.6.2). This implies that every national border contains a circuit (Theorem 2.5.9) and hence at least three circuit edges. Therefore, every country has at least three neighbors. From property 4 in the definition of regularity, these neighbors must be pairwise distinct. The existence of one country with at least three neighbors means that $\mathcal{L}$ has a minimum of four countries. □

Minimal criminals have particularly nice national borders. Their borders can have no unsightly protrusions.

Theorem 3.2.12

In a regular map that is a minimal criminal, the border of every country is a circuit. ∎

Proof Let $\mathcal{L}$ be a minimal criminal and L a country of $\mathcal{L}$. Since $\mathcal{L}$ has at least four countries, we can find a circuit $\mathcal{K} \subset \mathcal{G}_L$ (Theorem 2.5.9). We must show that $\mathcal{K} = \mathcal{G}_L$, from which we may, without loss of generality, assume that $L \subset I(\mathcal{K})$.

Assume that we can find an edge $B \in \mathcal{G}_L \setminus \mathcal{K}$. The fact that all points of B are arcwise accessible from within L implies that $\overset{\circ}{B} \subset I(\mathcal{K})$. Because of regularity, B is a circuit edge, and so we have another country L' to whose border B belongs (Lemma 2.6.2). By the arcwise accessibility of B from within L', it further follows that $L' \subset I(\mathcal{K})$. Thus $I(\mathcal{K})$ contains at least two countries.

A circuit consists of at least three edges. Now, all edges of $\mathcal{K}$ are borderlines of L. Due to the regularity condition, any two countries have at most one common borderline. Therefore, there must also exist a minimum of two, even a minimum of three, countries lying in $A(\mathcal{K})$. Moreover, none of these can have a common borderline with any country lying in $I(\mathcal{K})$ that is distinct from L.

We now construct two new maps $\mathcal{L}^i$ and $\mathcal{L}^a$. In the first map, we merge all countries lying in $I(\mathcal{K})$ with L to form one country L^i. In the second map $\mathcal{L}^a$, all countries contained in $A(\mathcal{K})$ are combined with L to form one country L^a. Each of these maps has fewer countries than $\mathcal{L}$ and hence has an admissible 4-coloring. We choose admissible 4-colorings φ^i for $\mathcal{L}^i$ and φ^a for $\mathcal{L}^a$ such that $\varphi^i(L^i) = \varphi^a(L^a)$. We now define $\varphi : \mathcal{M}_{\mathcal{L}} \to \{1, \ldots, 4\}$ as follows:

$$
\varphi(\tilde{L}) = \begin{cases} \varphi^a(L^a) & \text{for } \tilde{L} = L, \\ \varphi^a(\tilde{L}) & \text{for } L \neq \tilde{L} \subset I(\mathcal{K}), \\ \varphi^i(\tilde{L}) & \text{for } \tilde{L} \subset A(\mathcal{K}). \end{cases}
$$

In this way, we obtain an admissible 4-coloring for $\mathcal{L}$. This is in contradiction to our assumption. □

Corollary 3.2.13

In a map that is a minimal criminal, the degree of a vertex is equal to the number of countries that have this vertex as a border point. ∎

Proof Let $\mathcal{L}$ be a minimal criminal and $\mathbf{x}$ a vertex of $\mathcal{L}$. We pick an elementary neighborhood D of $\mathbf{x}$ and denote by $B_1, B_2, \ldots, B_d$ the circuit edges of $\mathcal{L}$ that are incident with $\mathbf{x}$. Label the sequence of circuit edges in such a way that the radii $D \cap B_t$ of D, for $t \in \{1, \ldots, d\}$, are cyclically ordered in a counterclockwise direction. Further, denote by $S_1, S_2, \ldots, S_d$ the sectors into which D is partitioned by the radii $D \cap B_t$, $t \in \{1, \ldots, d\}$. Label them in such a way that the sector S_t is incident with the radius $D \cap B_t$, again in a counterclockwise direction. Finally, for each $t \in \{1, \ldots, d\}$, we denote by L_t the country of $\mathcal{L}$ that contains the interior points of the sector S_t. Since the edges B_t are circuit edges, we have, in any case, that $L_t \neq L_{t+1}$ for $t \in \{1, \ldots, d-1\}$ and $L_d \neq L_1$ (Lemma 2.6.2). We must show that for $1 \leq t_1 < t_2 \leq d$, $L_{t_1} \neq L_{t_2}$ always holds.

We argue by contradiction. Without loss of generality, we assume that $L_1 = L_t$ for one t such that $2 < t < d$. Then $\mathbf{x}$ is incident with four distinct borderlines of L_1, namely, B_1, B_2, B_t, and B_{t+1}. By the previous theorem, the border of L_1 is a circuit. Hence, at most two borderlines of L_1 can be incident with one vertex. This provides the desired contradiction. $\qquad\square$

4
CHAPTER

From Topology to Combinatorics

As has been mentioned repeatedly in previous chapters, the Four-Color Problem can be formulated as a purely combinatorial statement without any reference to geometry or topology. This process of abstraction will now be presented.

4.1 Complete (Plane) Graphs

We will be systematically examining a special kind of map, and we shall do this more accurately than is possible within the context of examples. The results are of a graph-theoretical nature. Therefore, graph-theoretical terminology will be used primarily. In the process, we will also be considering graphs with isolated vertices. However, multiple edges and loops will still be excluded from the discussion. If $G = (E, \mathcal{L})$ is a graph, then we denote by N_G the neutrality set of the corresponding map together with its isolated vertices. In other words,

$$N_G = N_{\mathcal{L}} \cup E.$$

Definition 4.1.1
A graph is said to be *complete* if for each pair of vertices, there is an edge joining them.

There exist complete (plane!) graphs only with two, three, or four vertices.

Here are the typical examples.

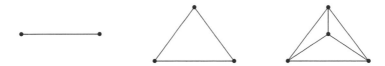

Theorem 4.1.2
There exist no complete graphs with five vertices. ■

Proof Let $\mathbf{x}_1, \ldots, \mathbf{x}_5$ be five distinct points in the plane. For each pair of numbers $i, j \in \{1, \ldots, 5\}$ such that $i < j$, let B_{ij} be an arc with the given $\mathbf{x}_i$ and $\mathbf{x}_j$ as end points but that does not contain any of the other given points. Altogether there are ten arcs in question. Seven of them have either $\mathbf{x}_1$ or $\mathbf{x}_5$ (or both) as an end point. Assume that each pair of these seven arcs has no interior points in common. It will be shown that at least one of the three remaining arcs has an interior point in common with one of the previous seven. By concatenation, we obtain three arcs $B_i = B_{1i} \cup B_{i5}$, $i = 2, 3, 4$, which, like B_{15}, join the points $\mathbf{x}_1$ and $\mathbf{x}_5$ but which have neither common interior points within themselves nor common interior points with B_{15}. Of these three, there is exactly one, let's say B_3, with the property that one of the two remaining arcs lies in the interior domain while the other lies in the exterior domain of the closed Jordan curve $K = B_{15} \cup B_3$ (Theorem 2.2.11). Therefore, the arc B_{24} that joins the interior point $\mathbf{x}_2$ of B_2 with the interior point $\mathbf{x}_4$ of B_4 must have at least one interior point $\mathbf{y}$ in common with K (part 3 of the Jordan curve theorem 2.2.5). This means that B_{24} has at least one interior point in common with B_{15}, with B_{13} or with B_{35}. Since B_{24} is not incident with the vertices $\mathbf{x}_1, \mathbf{x}_3$, or $\mathbf{x}_5$, then $\mathbf{y}$ must be an interior point of B_{15}, B_{13}, or B_{35}.   □

If two distinct points of the plane are given, we can then choose an arbitrary arc joining them, for instance, the line segment between them. In this way, we obtain a complete graph with two vertices.

A complete graph with three vertices will also be called a *(curvilinear) triangle*.[1] Three distinct points of the plane are always vertices of a curvilinear triangle. If the three points do not lie in a straight line, they form the (rectilinear) triangle shown in the diagram preceding Theorem 4.1.2. The following illustration gives one possible construction of a triangle in the case that the points do lie in a straight line.

Moreover:

Theorem 4.1.3
Every graph with at most three (including isolated) vertices can be completed to form a curvilinear triangle. ∎

Proof In principle, there are numerous cases to consider. We will limit ourselves to the least trivial of them, that is, a graph with three vertices and one edge (and, consequently, one isolated vertex). In this case, we have three points $\mathbf{x}_1$, $\mathbf{x}_2$, and $\mathbf{x}_3$ in the plane and one arc B_{12} joining $\mathbf{x}_1$ and $\mathbf{x}_2$ that does not contain $\mathbf{x}_3$. Now, $\mathbf{x}_3$ lies in the only country of the map formed from the arc B_{12} (Example 2.4.2). Therefore, it can be joined to the vertex $\mathbf{x}_1$ by an arc B_{13} in such a way that the set $\{B_{12}, B_{13}\}$ also forms a map (Corollary 2.3.12). This map still has no circuits. Therefore, it too has only one country L (Theorem 2.4.4). As the points $\mathbf{x}_2$ and $\mathbf{x}_3$ are border points of L, they can be joined by an arc B_{23} (Lemma 2.5.2, part b) in such a way that the pair

$$(\{\mathbf{x}_1, \mathbf{x}_2, \mathbf{x}_3\}, \{B_{12}, B_{13}, B_{23}\})$$

forms a curvilinear triangle. This is the required completion of the given graph. □

There are equally many complete graphs with four vertices. They will be called *complete 4-gons*. Besides four vertices, they have six

---

[1]The notion of a *triangle* has already been defined (see page 69). However, in order to avoid an excess of "mental gymnastics" in regard to these concepts, we tolerate a certain amount of ambiguity. From the context, it will always be clear which meaning of *triangle* is being considered.

edges. For each edge B, there exists exactly one *counteredge*, which is an edge B' having no common vertices with B. Three pairs of counteredges arise in this way. The following theorem is an analogue of the previous theorem.

Theorem 4.1.4

Every graph with at most four (including isolated) vertices can be extended to form a complete 4-gon.   ■

Proof   In this theorem as well, several different cases must be examined. As long as the graph, up to its completion, is circuit-free, we can proceed as in the previous proof. One must only be convinced that in the choice of the required arcs, one can avoid isolated points.

Consequently, we may begin with a graph that consists of one curvilinear triangle with vertices $\mathbf{x}_1$, $\mathbf{x}_2$, $\mathbf{x}_3$ and one additional isolated vertex $\mathbf{x}_4$. The edges that join the vertices $\mathbf{x}_i$ and $\mathbf{x}_j$, $1 \leq i < j \leq 3$, will be denoted by B_{ij}. The map $\mathcal{L} = \{B_{12}, B_{23}, B_{13}\}$ is a circuit and has two countries (Example 2.4.6). In one of them lies $\mathbf{x}_4$. This country will be denoted by L. By using the stereographic projection, we can assume that L is the bounded country of the map $\mathcal{L}$. We pick an arc B_{14}, joining $\mathbf{x}_1$ to $\mathbf{x}_4$, whose interior points lie entirely in L (Corollary 2.3.12). By adding B_{14} to $\mathcal{L}$, a map $\mathcal{L}'$ is formed having B_{14} as a final edge. Now, $L' = L \setminus B_{14}$ is a country of $\mathcal{L}'$ (Case 2, page 82) with $\mathbf{x}_4$ and $\mathbf{x}_2$ as border points. Therefore, we can find an arc B_{24} joining $\mathbf{x}_4$ to $\mathbf{x}_2$ all of whose interior points belong to L' (again Lemma 2.5.2, part b). The set-theoretic union $K = B_{13} \cup B_{14} \cup B_{24} \cup B_{23}$ is a closed Jordan curve whose exterior domain, by assumption, contains the interior points of B_{12}. By the Schoenflies theorem (Theorem 2.2.7), we can further assume that K is the unit circle. Then the chord B_{34}, which joins the points $\mathbf{x}_4$ and $\mathbf{x}_3$, extends the already constructed configuration to the desired complete 4-gon.   □

Remark: Anyone reading the above proof may ask why we didn't apply the Schoenflies theorem directly to the neutrality set of the map $\mathcal{L}$. This is because the construction presented here lends itself to a simple modification if in addition to the edges of $\mathcal{L}$, the edge B_{14} or the edges B_{14} and B_{24} are given.

However, complete graphs with four vertices are subject to some restrictions. First, we note that:

Lemma 4.1.5
Interior points of two counteredges of a complete 4-gon can be joined only by arcs that have a point in common with at least one other edge. ∎

Proof Let G be a complete 4-gon with the four vertices $\mathbf{x}_i$, $i = 1, 2, 3, 4$. For all i, j such that $i < j$, let B_{ij} denote the edge joining the vertices $\mathbf{x}_i$ and $\mathbf{x}_j$. In addition, let points $\mathbf{y} \in \overset{\circ}{B}_{13}$ and $\mathbf{z} \in \overset{\circ}{B}_{24}$ be given. We denote by K the closed Jordan curve $B_{12} \cup B_{23} \cup B_{34} \cup B_{14}$. We want to show that one of the points $\mathbf{y}, \mathbf{z}$ lies in the interior domain of K and the other lies in the exterior domain of K. With this, the result has been proved.

First, we assume that $\mathbf{y} \in I(K)$. Then all the interior points of B_{13} lie in the interior domain of K. Because the pairs $(\mathbf{x}_1, \mathbf{x}_3)$ and $(\mathbf{x}_2, \mathbf{x}_4)$ are separated in K, it follows that $\overset{\circ}{B}_{24} \subset A(K)$ (Corollary 2.2.12) and therefore $\mathbf{z} \in A(K)$.

The second case follows analogously. □

An interesting consequence of this fact is the unsolvability of the famous "supply problem." Three houses must be linked with cables for three utility services (electricity, gas, and water). These cables cannot overlap but must all lie within a plane (an unnecessary restriction in practical terms). Dudeney, already in 1917, wrote the following about this problem: "... *as old as the hills ... much older than electrical lights or even gas, but a different guise brings it up to date*" [DUDENEY 1958, page 73]. The problem is solvable only by an impermissible trick [DUDENEY 1958, page 200].

Corollary 4.1.6
Let $\mathbf{x}_1$, $\mathbf{x}_2$, $\mathbf{x}_3$, $\mathbf{y}_1$, $\mathbf{y}_2$, $\mathbf{y}_3$ be six distinct points in the plane. Suppose that each of the points $\mathbf{x}_i$, $i = 1, 2, 3$, is joined to each of the points $\mathbf{y}_j$, $j = 1, 2, 3$, by an arc B_{ij} that except for its end points contains none of the given points. Then at least two out of the nine arcs B_{ij} in total have a common interior point. ∎

Proof We can assume that the eight arcs B_{ij}, where $ij \neq 33$, pairwise have no common interior points. Then we create a complete 4-gon from the points $\mathbf{x}_1$, $\mathbf{x}_2$, $\mathbf{y}_1$, $\mathbf{y}_2$, together with the arcs B_{11}, B_{12}, B_{21}, B_{22}, $B_{13} \cup B_{23}$, $B_{31} \cup B_{32}$. The points $\mathbf{x}_3$ and $\mathbf{y}_3$ are interior points of counteredges of this 4-gon. Therefore, the arc B_{33} must intersect one of the other arcs B_{ij} at an interior point. □

The second restriction on complete 4-gons concerns the notion that de Morgan, albeit mistakenly (see page 11), saw as the core problem of the Four-Color Theorem.

Theorem 4.1.7

A complete graph with four vertices has exactly one vertex that lies in the interior domain of the closed Jordan curve formed from the three edges not incident with it. ■

Proof Let G be a complete 4-gon with the four vertices x_i, $i = 1, 2, 3, 4$. For all i, j where $i < j$, let B_{ij} denote the edge joining the vertices x_i and x_j. The interior points of exactly one of the three arcs B_{12}, $B_{13} \cup B_{23}$, or $B_{14} \cup B_{24}$ lies in the interior domain of the closed Jordan curve formed by the other two (Theorem 2.2.11). If this is not the case for the arc B_{12}, then either x_3 or x_4 lies in the interior domain of the closed Jordan curve formed by the edges not incident with the vertex. In either case, the other vertices can be joined to points at an arbitrary distance away by arcs that do not intersect N_G. Hence, these vertices cannot lie in the interior domain of any closed Jordan curve formed from the edges of G.

One case remains to be considered. This is the case in which the interior points of B_{12} lie in the interior domain of the closed Jordan curve $K = B_{13} \cup B_{23} \cup B_{24} \cup B_{14}$. From this, it follows that $\overset{\circ}{B}_{34} \subset A(K)$ (Corollary 2.2.12). Now, the interior points of exactly one of the three arcs B_{34}, $B_{13} \cup B_{14}$, or $B_{23} \cup B_{24}$ must lie in the interior domain of the closed Jordan curve formed by the other two. This cannot be the case for the arc B_{34}, and so one of the points x_1 or x_2 is the enclosed point. □

From an abstract point of view, this theorem has to do with the properties of the ordering of the plane—the way it was formulated by Pasch and Hilbert a few decades following de Morgan. The precise relationship between de Morgan's basic approach and the so-called Pasch axiom has not to this day been fully understood.

4.2   The Wagner and Fáry Theorem

In the context of the Four-Color Problem, Wagner, in 1936, showed that the graphs in the plane that he utilized could be composed solely of edges that were straight line segments. In 1947, Fáry, independently of Wagner, formulated and proved the same assertion for arbitrary graphs without multiple edges and loops. He remarked that this had been conjectured by Tibor Szele in a discussion on the characterization of planar graphs (Definition 5.2.4) that had been given by Kuratowski in 1930.

In this section we will prove a somewhat more general version of the Wagner and Fáry theorem. This theorem brings the final liberation from the house of horrors of arbitrary simple curves, and it releases the Four-Color Theorem from the jaws of general topology. It enables one in the investigation of maps to limit oneself to those maps whose edges are line segments. From the standpoint of the mathematical methodology under consideration, this theorem belongs to the fundamental tenets of graph theory. Thus, in this section we will also, for the most part, be using standard graph-theoretical terminology. For instance, a *face of a graph* refers to a country of the associated map. Its *unbounded face*, sometimes called its *infinite face*, will correspond to the unbounded country. Its *bounded faces*, sometimes referred to as its *finite faces*, will denote the bounded countries.

Λ particular kind of graph will be needed in order to carry out the proof. We now begin our study of them.

Saturated Graphs

Definition 4.2.1
A plane graph is said to be *saturated* if it is not a proper subgraph of another plane graph having the same vertex set.

In practical terms, a plane graph is saturated if no new edges can be added to the graph without enlarging the vertex set. Instead of "saturated," one sometimes says "maximally plane" or "triangulated."

The reason for the last of these terms will be made clear in what follows.

Complete graphs are saturated. The converse, as is illustrated by the following diagram, is, in general, false.

One sees immediately that were the graph represented above not saturated, then it would be a complete graph with five vertices—but such a graph cannot exist (Theorem 4.1.2).

Lemma 4.2.2

In a saturated graph, any two vertices that belong to the boundary of one and the same face are joined by an edge. ∎

Proof Two points on the boundary of a face can always be joined by an arc that, up to its end points, lies entirely in the interior of this face (Lemma 2.5.2). □

Lemma 4.2.3

A saturated graph is connected. ∎

Proof Let G be a graph whose neutrality set N_G has several different components. Then there must exist a face whose boundary is not connected (Theorem 2.5.10). Every boundary component of such a face contains vertices of G (Theorem 2.5.4). Two vertices in different boundary components of a face cannot, however, be joined by an edge. □

The faces into which a saturated graph in the plane is subdivided have particularly simple boundaries.

Theorem 4.2.4

A graph with at least three vertices is saturated if and only if the borders of all its faces are (curvilinear) triangles. ∎

Proof If a graph is not saturated, then it must have two vertices that are not the end points of one and the same edge but that can be joined by an arc having no interior points in common with the neutrality set. The existence of such an arc implies that both of these vertices belong to the boundary of one and the same face. If the border of a face, however, is a curvilinear triangle, then any two vertices in its boundary are joined by an edge. Thus, the stated condition is sufficient. To prove that the condition is necessary is somewhat lengthy and requires several steps.

A saturated graph with three vertices is a curvilinear triangle (Theorem 4.1.3). Moreover, the set-theoretic union of its edges is a closed Jordan curve. By the Jordan curve theorem (Theorem 2.2.5), there exist exactly two faces, and all three vertices belong to their common boundary.

A saturated graph with four vertices is a complete 4-gon (Theorem 4.1.4). Now, one of the four vertices lies in the interior domain of the closed Jordan curve formed by the edges not incident with it (Theorem 4.1.7). In any case, the boundary of the unique unbounded (or infinite) face contains exactly three vertices. By a double application of the stereographic projection, we can now transform each face into the infinite face. In doing so, we alter neither the completeness of the graph nor the number of vertices in the boundary of a face. From this, the claim follows for saturated graphs with four vertices.

Let G be a saturated graph with more than four vertices. If it were circuit-free, then there would exist three vertices in the boundary of its only face (Theorem 2.4.4) that pairwise would not be joined by edges of G. By Lemma 4.2.2, this cannot happen. Therefore, there exists at least one circuit and with it at least two faces. It then follows that the boundary of each face contains a circuit (Theorem 2.5.9) and that at least three vertices belong to each of these circuits.

Finally, we must show that the boundary of a face of G can contain at most three vertices. To do this, we assume that G has a face L whose border $B(L)$ contains more than three vertices. Since every two vertices in $B(L)$ can be joined by an edge (Lemma 4.2.2), a complete 4-gon G' is formed from four distinct chosen vertices in $B(L)$, together with the six edges of G joining them. The face L has no point in common with $N_{G'}$. Hence, because L is path connected, L lies entirely in a face L' of G'. Because G' is a complete 4-gon, the boundary $B(L')$ contains exactly three vertices. Therefore, one of the chosen vertices does not belong to $B(L')$. Since $L \subset L'$, this vertex also cannot belong to $B(L)$. This provides the desired contradiction.   □

Out of this theorem another interesting property of saturated graphs emerges.

Corollary 4.2.5

A saturated graph with more than two faces is regular.   ∎

Proof Let $G = (E, \mathcal{L})$ be a saturated graph with more than two faces. We must verify the properties of regularity (Definition 3.2.8) for $\mathcal{L}$. A graph with more than two faces certainly has edges. Therefore, $\mathcal{L} \neq \emptyset$. Connectedness we have already established (Lemma 4.2.3). By the previous theorem, in a saturated graph only triangles crop up as borders of faces. Since each side of a triangle is always a circuit edge, no bridges or final edges can exist in G.

If two triangles were to have exactly two sides in common, their third sides would be distinct from each other and would have the same end points. However, in the definition of (plane) graphs, multiple edges are not permitted. From the previous theorem, it also follows that no two distinct faces can have exactly two common borderlines.

Again, by the previous theorem, because no border of a face contains more than three edges, there remains one more case to consider. This is the case when a graph has two faces in which exactly three, and hence all borderlines are common. This is certainly true in a saturated graph that is a triangle. However, in this case there can only be two faces, a situation that has been excluded from the outset.

In general, let L_1, L_2 be two faces of G with three common borderlines. These three edges form a circuit $\mathcal{K}$ that separates the two faces (Theorem 2.6.6 and Corollary 2.6.7). Without loss of generality, we can assume that L_1 lies in $I(\mathcal{K})$. We claim that in fact, $L_1 = I(\mathcal{K})$. If there were to exist a point $\mathbf{x} \in I(\mathcal{K}) \backslash L_1$, then this $\mathbf{x}$ could be joined to a point $\mathbf{y} \in L_1$ by an arc lying entirely in $I(\mathcal{K})$. Such an arc must contain a border point of L_1 — a situation that is not possible because $L_1 \cap N_{\mathcal{K}} = \emptyset$. Analogously, it follows that $L_2 = A(\mathcal{K})$. Therefore, G has only two faces L_1 and L_2. This is in contradiction to the assumption that G has more than two faces.   □

Next, we must make available a device that is actually not of a typical graph-theoretical flavor.

Digression: Star-Shaped Polygons

Definition 4.2.6

A point $\mathbf{z}$ is a *star point* of a polygon P if $\mathbf{z}$ belongs to the interior domain $I(P)$ of P and if for each point $\mathbf{x} \in P$, the line segment from $\mathbf{z}$ to $\mathbf{x}$, excluding $\mathbf{x}$ itself, contains only points of $I(P)$. A polygon is said to be *star-shaped* if it has a star point.

This foundational concept of a star point, due to [WAGNER 1936], is more restricted in its definition than in general topology, where $\mathbf{z}$ itself and also subsegments of the line segment $[\mathbf{z}, \mathbf{x}]$ are permitted to belong to P.[2] This notion will be needed in this form in order to ensure the validity of the following assertion.

Lemma 4.2.7

The set of star points of a polygon is an open set. ■

Proof Let P be a polygon consisting of the abutted line segments $S_1, S_2, \ldots, S_n$. For each i, we pick a point $\mathbf{x}_i \in \overset{\circ}{S}_i$ and an elementary neighborhood D_i of $\mathbf{x}_i$. Each disk D_i will be subdivided into two semicircular disks $D_i^+ = D_i \cap I(P)$ and $D_i^- = D_i \cap A(P)$ by the given line segments S_i. We then denote by H_i the open half-plane[3] that encompasses the semicircular disk D_i^+ and whose boundary contains the line segment S_i. We now claim: *The set Z of star points of P is the intersection of all the half-planes H_i, $1 \le i \le n$.* Hence, Z is itself an open set. We must therefore only show that

$$Z = \bigcap_{i=1}^{n} H_i .$$

---

[2]See, for example, [CIGLER—REICHEL 1987, page 159].
[3]A half-plane is a subset of the plane that consists of one side of a straight-line axis. It is closed or open according to whether the bordering axis belongs to it or not.

In the first place, let a star point $\mathbf{z}$ of P be given. From the definition, the line segment $S_i' = [\mathbf{z}, \mathbf{x}_i]$, except for end point $\mathbf{x}_i$, lies entirely in $I(P)$. Now, S_i' must have points in common with D_i^+ and therefore, again except for the end point $\mathbf{x}_i$, must lie entirely in H_i. It follows that $\mathbf{z} \in H_i$. Conversely, suppose a point $\mathbf{z} \in \bigcap_{i=1}^n H_i$ is given. We consider the line segment S from $\mathbf{z}$ to an arbitrary point $\mathbf{x} \in P$. Then $\mathbf{x} \in S_i$ for (at least) one $i \in \{1, \dots, n\}$, and $S \subset H_i \cup \{\mathbf{x}\}$. Since $\mathbf{z} \in H_i$, it further guarantees the existence of subsegments of S with $\mathbf{x}$ as an end point whose interior points completely lie in $I(P)$. Let S' denote the subsegment of S that is maximal with respect to this property. Let $\mathbf{y}$ denote the other end point of S'. We must show that $\mathbf{y} = \mathbf{z}$. We argue by contradiction. If $\mathbf{y} \in \overset{\circ}{S}$, then $\mathbf{y} \in S_j$ for some $j \neq i$, and $\overset{\circ}{S'} \subset H_j$. Hence $\mathbf{z} \notin H_j$. This, of course, cannot happen. $\square$

The above proof also establishes that:

Corollary 4.2.8
The polygon P is star-shaped if and only if the half-planes H_i, $i = 1, \dots, n$, have a nonempty intersection. In other words,

$$\bigcap_{i=1}^n H_i \neq \emptyset.$$

    ∎

With this result at our disposal, the important partitioning property of star-shaped polygons is now easy to prove.

Theorem 4.2.9
Suppose that P is a star-shaped polygon and that $\mathbf{z}$ is a star point of P. Let $\mathbf{x}_1$ and $\mathbf{x}_2$ be two distinct points of P. Let S denote one of the polygonal arcs into which P is subdivided by the points $\mathbf{x}_1$ and $\mathbf{x}_2$. Then the polygon $P' = [\mathbf{z}, \mathbf{x}_1] \cup S \cup [\mathbf{x}_2, \mathbf{z}]$ is star-shaped.   ∎

Proof For $j = 1, 2$, let S_j' be the line segment belonging to S whose end point is $\mathbf{x}_j$. Let H_j' denote the open half-plane bordered by the straight line through the points $\mathbf{z}$ and $\mathbf{x}_j$ that contains the line segment S_j' (except for the end point $\mathbf{x}_j$). Because $\mathbf{x}_1 \neq \mathbf{x}_2$, $H_1' \cap H_2'$ is a nonempty open unbounded sector of the plane. In what follows, we will also be using the notation adopted in the proof of Lemma 4.2.7. Thus the line segments S_j' are subsegments of certain segments S_i of

S. The corresponding half-planes H_i can therefore be used to determine the set of the star points of P. Furthermore, we can establish that for the set Z' of the star points of P', the following is true:

$$Z' \supset H_1' \cap H_2' \cap \bigcap_{i=1}^{n} H_i = H_1' \cap H_2' \cap Z \,.$$

Because Z is open (Lemma 4.2.7), we can find an open disk U centered at $\mathbf{z}$ that is entirely contained in Z. As $\mathbf{z}$ belongs to the boundary of not only H_1' but also H_2', $H_1' \cap H_2' \cap U$ is a nonempty open circular sector contained in Z'. Hence, $Z' \neq \emptyset$ and P' is star-shaped (Corollary 4.2.8). □

Remark (Rectilinear): 3-, 4- and 5-gons are always star-shaped. The proof of the existence of a 6-gon that is not star-shaped and its construction will be left to the reader.

The Theorem

Definition 4.2.10
A *line graph* is a graph whose edges are all line segments.

Theorem 4.2.11 (Wagner and Fáry Theorem)
Every graph can be transformed into a line graph by a homeomorphism of the plane into itself. ■

Proof Every graph with at most three vertices can be completed to form a curvilinear triangle (Theorem 4.1.3). By the Schoenflies theorem (Theorem 2.2.7), such a curvilinear triangle can be transformed into a rectilinear triangle by a homeomorphism of the plane into itself. Therefore, we need only consider graphs with more than three vertices.

An arbitrary graph can be extended to a saturated graph by the addition of edges. Any homeomorphism of the plane into itself that transforms such an extended graph to a line graph will do the same to the original graph. For this reason, we need only deal with saturated graphs.

Let $\mathbf{G}$ be a saturated graph with v vertices, $v \geq 4$. The first step in the construction of a homeomorphism of the desired type consists in the clever choice of a series of subgraphs G_k of G where $k = 3, 4, \ldots, v$. These graphs G_k should fulfill the following conditions:

1. G_3 consists of vertices and edges that belong to the boundary of the infinite face of G. G_3, therefore, is a curvilinear triangle, whose vertices we will denote by x_1, x_2, x_3.

2. $G_v = G$.

3. If the end points of an edge B of G are vertices of G_k, $k = 3, 4, \ldots, v$, then B belongs to G_k.

4. For $k = 4, \ldots, v$, G_k has exactly one vertex more than G_{k-1}. This vertex will be denoted by x_k. This means that every subgraph G_k has exactly k vertices.

5. Every vertex x_k, $k = 4, \ldots, v$, is the end point of at least two edges of G whose other end points are vertices of G_{k-1}.

We obtain such a series of subgraphs of G by constructing them inductively. The first step in the induction process is clear: G_3 is formed from the first stipulated condition. Now let $G_{k-1}, k = 4, \ldots, v$, be given. The difficulty in the induction step lies in the discovery of a vertex x_k that fulfills the fourth and fifth conditions. The third condition can then be used to construct the requisite subgraph G_k in the following way. G_{k-1} is extended to include the vertex x_k and all edges of G that join a vertex of G_{k-1} to x_k. Because G_{k-1} has precisely $k-1$ vertices and $k-1 < v$, we can certainly find a vertex x of G that does not belong to G_{k-1}. Therefore, x must lie in a face L' of G_{k-1}. We choose an edge B of G_{k-1} that belongs to the border of L'. In the transition from G_{k-1} to G, L' will be further subdivided into faces. We find a face L of G that is contained in L' and to whose border the edge B also belongs. Because G is saturated, the border $\mathcal{G}_L$ of L is a curvilinear triangle (Theorem 4.2.4). Two vertices of $\mathcal{G}_L$, namely, the end points of B, belong to G_{k-1}, but the third does not. Note that if all three vertices of $\mathcal{G}_L$ were to belong to G_{k-1}, then L would also be a face of G_{k-1}, by the third of the conditions specified above. Therefore, L would equal L'. This is not possible, since $x \in L' \setminus L$. Because the edges in $\mathcal{G}_L$ belong to G, the vertex of $\mathcal{G}_L$ that does not belong to G_{k-1} is joined by two distinct edges of G to vertices of G_{k-1}, namely, the end points of B. This vertex we choose for x_k. In this way, the required properties are fulfilled.

After having constructed a sequence G_k of subgraphs of $\mathbf{G}$ with the specified properties, we are now in a position to define the desired homeomorphism, also inductively. We obtain a series of homeomorphisms $h_k : \mathbb{R}^2 \rightarrow \mathbb{R}^2$, $k = 3, 4, \ldots, v$, that transform the respective G_k's into line graphs G'_k whose given faces will be bordered by star-shaped(!) polygons. In this way, by the construction of the homeomorphism h_v, the theorem is proved.

In order to realize this, we start with a homeomorphism h_3 that transforms the curvilinear triangle G_3 into a rectilinear triangle. The existence of h_3 is guaranteed by the Schoenflies theorem (Theorem 2.2.7). Moreover, a rectilinear triangle is a star-shaped polygon.

Suppose then that h_{k-1} has been constructed. We denote by G''_k the graph into which G_k has been transformed by h_{k-1}. We set $\mathbf{y}_j = h_{k-1}(\mathbf{x}_j)$ for $j = 1, 2, \ldots, k - 1$. The vertex $\mathbf{y}''_k = h_{k-1}(\mathbf{x}_k)$ of G''_k lies in a face L' of G'_{k-1} whose border $B(L')$ is a star-shaped polygon. This vertex is joined in G''_k to a few of the vertices $\mathbf{y}_j$ by the edges B_j. We denote by J the subset of the indices $1, 2, \ldots, k - 1$ for which this applies. The face L' is subdivided into finitely many faces L''_1, $L''_2, \ldots, L''_p$ in G''_k. These latter faces are bounded by closed Jordan curves, each consisting of a polygonal arc in $B(L')$ and two of the edges B_j for $j \in J$. We choose a star point $\mathbf{y}_k$ of the polygon $B(L')$ and obtain the desired line graph G'_k by extending G'_{k-1} as follows. Add the point $\mathbf{y}_k$ to the set of vertices of G'_{k-1} and add the line segments $[\mathbf{y}_j, \mathbf{y}_k]$ for all $j \in J$ as edges. The face L' will, by this process, be subdivided into finitely many faces $L'_1, L'_2, \ldots, L'_p$ of G'_k having star-shaped borders (Theorem 4.2.9). We can number them in such a way so that

$$B(L'_r) \cap B(L') = B(L''_r) \cap B(L')$$

holds for $r = 1, 2, \ldots, p$.

The remaining faces of G_{k-1} are not altered in the transition to G_k. By the induction hypothesis, they will all have star-shaped polygons as borders. Therefore, all faces of G_k have star-shaped borders. Now we define a homeomorphism $h'_k : \mathbb{R}^2 \rightarrow \mathbb{R}^2$ by the following steps:

- The points exterior to L' are mapped to themselves.

- The edges $B_j, j \in J$, are mapped to the line segments $[\mathbf{y}_j, \mathbf{y}_k]$.

- We extend the thus induced homeomorphisms from $B(L_r'') \rightarrow$ $B(L_r')$, $r = 1, 2, \ldots, p$, with the help of the Schoenflies theorem (Theorem 2.2.7) to the entire plane. The faces L_r'' are mapped homeomorphically to the faces L_r' by restricting the extensions to the interior faces.

Finally, we set $h_k = h_k' \circ h_{k-1}$. $\qquad\qquad\qquad\qquad$ □

Remark: The proof of the Wagner and Fáry theorem presented here relies upon the foundational arguments of Wagner [WAGNER 1936] (see also the presentation in [WAGNER-BODENDIEK 1989, Theorem 2.1]). Fáry's method of proof [FÁRY 1947] is perhaps somewhat shorter because in his proof the comprehensive discussion of star-shaped polygons is not necessary. However, his method is more indirect. One who can construct the desired line graphs step by step understands the proof better than one who has not done so (see also [ORE 1967, Theorem 1.3.3]).

4.3   The Euler Polyhedral Formula

In the combinatorial treatment of maps and graph-theoretical problems, counting arguments play a crucial role. For that, we will be using the following notation:

$v_{\mathcal{L}}$ denotes the number of vertices,

$e_{\mathcal{L}}$ denotes the number of edges,

$f_{\mathcal{L}}$ represents the number of countries (f meaning "face"), and

$z_{\mathcal{L}}$ signifies the number of components

of a map $\mathcal{L}$. First of all, we will consider how these numbers change when one adds an edge to a given map (Section 2.7).

Lemma 4.3.1
Suppose $\mathcal{L}'$ is the map formed from the map $\mathcal{L}$ by the addition of an edge B. The following table calculates the numbers of vertices, edges, countries, and components of $\mathcal{L}'$ in comparison to the corresponding numbers for the map $\mathcal{L}$.

Type of edge B (as an edge in $\mathcal{L}'$)	$v_{\mathcal{L}'}$	$e_{\mathcal{L}'}$	$f_{\mathcal{L}'}$	$z_{\mathcal{L}'}$
final edge with two final vertices	$v_{\mathcal{L}} + 2$	$e_{\mathcal{L}} + 1$	$f_{\mathcal{L}}$	$z_{\mathcal{L}} + 1$
final edge with one final vertex	$v_{\mathcal{L}} + 1$	$e_{\mathcal{L}} + 1$	$f_{\mathcal{L}}$	$z_{\mathcal{L}}$
bridge	$v_{\mathcal{L}}$	$e_{\mathcal{L}} + 1$	$f_{\mathcal{L}}$	$z_{\mathcal{L}} - 1$
circuit edge	$v_{\mathcal{L}}$	$e_{\mathcal{L}} + 1$	$f_{\mathcal{L}} + 1$	$z_{\mathcal{L}}$

Proof The claim in regard to $e_{\mathcal{L}'}$ follows immediately from the construction of $\mathcal{L}'$. The others must be considered according to the type of the edge B:

1. If B is a final edge in $\mathcal{L}'$ with two final vertices, then no end point of B is the end point of an edge in $\mathcal{L}$. In this case, $\{B\}$ is a component of $\mathcal{L}'$. In the extension from one map to the other, the two end points of B become additional vertices and $\{B\}$ an additional component. From an earlier discussion on the extension of maps, it follows that the number of countries remains the same (Case 1, page 81).

2. If B is a final edge in $\mathcal{L}'$ with one final vertex, then exactly one end point of B is a final vertex of $\mathcal{L}'$. This end point is an additional vertex in the extension. Therefore, $v_{\mathcal{L}'} = v_{\mathcal{L}} + 1$. The number of countries in this case as well does not change (Case 2, page 82). Two vertices of $\mathcal{L}'$ that can be joined by an arc consisting of edges of $\mathcal{L}$ can also be joined by an arc consisting of edges of $\mathcal{L}'$. Hence, $z_{\mathcal{L}'} \leq z_{\mathcal{L}}$. If $\tilde{B}$ is an arc formed from edges in $\mathcal{L}'$ that joins two vertices of $\mathcal{L}$, then $\tilde{B}$ can (up to the end points) contain only vertices that are not final vertices of $\mathcal{L}'$. Therefore, B cannot be a subset of $\tilde{B}$. In other words, $\tilde{B}$ is an arc consisting of edges of $\mathcal{L}$. It follows that $z_{\mathcal{L}'} \geq z_{\mathcal{L}}$. Consequently, $z_{\mathcal{L}'} = z_{\mathcal{L}}$.

3. If B is a bridge in $\mathcal{L}'$, then neither vertex of B is a final vertex of $\mathcal{L}'$. Both vertices are therefore end points of edges in $\mathcal{L}$. Hence, the number of vertices does not change by the extension. By the same token, the number of countries also remains the same (Case 3, page 82). Both end points of B cannot, however, be joined by an arc consisting only of edges in $\mathcal{L}$, because the edges that would form such an arc, together with B, would form a circuit in $\mathcal{L}'$. This would imply that B would be a circuit edge in $\mathcal{L}'$. Therefore, the end points of B belong to two different components $\mathcal{L}_1$, $\mathcal{L}_2$ of $\mathcal{L}$. The components of $\mathcal{L}$ that do not contain the end points

of B as vertices remain unchanged in the extension from one map to the other. However, each vertex of $\mathcal{L}_1$ can be joined to each vertex of $\mathcal{L}_2$ by an arc formed from edges in $\mathcal{L}'$. To see this, let a vertex of $\mathcal{L}_1$ be given. We can then find an arc consisting of edges in $\mathcal{L}_1$ that joins this vertex to an end point of B. This arc we extend to include B itself and link it to an arc consisting of edges of $\mathcal{L}_2$ that joins the second end point of B to an arbitrary vertex in $\mathcal{L}_2$. Consequently, in the extension, both of the components $\mathcal{L}_1$, $\mathcal{L}_2$ of $\mathcal{L}$ are merged into a single component of $\mathcal{L}'$. This lowers the number of components by one.

4. If B is a circuit edge in $\mathcal{L}'$, then both end points of B are again end points of edges in $\mathcal{L}$. Hence, the number of vertices in the extension remains the same. The country of $\mathcal{L}$ containing $\overset{\circ}{B}$ will be subdivided into two countries. All other countries remain unchanged (Case 3, possibility 2, page 82). This means that $f_{\mathcal{L}'} = f_{\mathcal{L}} + 1$. Since both end points of B, from the definition of circuit edge, can be joined by an arc consisting of edges of $\mathcal{L}$, the number of components also remains the same.   □

Since there exist only the four possibilities specified above for an edge of a map, we obtain the following important equation.

Corollary 4.3.2
Suppose the map $\mathcal{L}'$ is formed from the map $\mathcal{L}$ by the addition of an edge. Then the following equation holds:

$$v_{\mathcal{L}'} - e_{\mathcal{L}'} + f_{\mathcal{L}'} - z_{\mathcal{L}'} = v_{\mathcal{L}} - e_{\mathcal{L}} + f_{\mathcal{L}} - z_{\mathcal{L}}. \qquad (4.3.1)$$
■

The addition of finitely many edges is nothing more than a finite repetition of the process of adding a single edge. For this reason, equation 4.3.1 also holds if the map $\mathcal{L}'$ is created from $\mathcal{L}$ by the inclusion of finitely many edges. However, since every map can be formed from the empty map by the addition of finitely many edges, the number $v_{\mathcal{L}} - e_{\mathcal{L}} + f_{\mathcal{L}} - z_{\mathcal{L}}$ is the same for each map. In other words, it is precisely the number that holds for the empty map as well. The empty map has no vertices, no edges, and no components, but it has exactly one country (Example 2.4.1). Hence, $v_{\emptyset} - e_{\emptyset} + f_{\emptyset} - z_{\emptyset} = 1$. With this, we obtain the formula that was first published by Augustin-Louis Cauchy in 1813.

Theorem 4.3.3

For each map $\mathcal{L}$, the following formula holds:

$$v_{\mathcal{L}} - e_{\mathcal{L}} + f_{\mathcal{L}} - z_{\mathcal{L}} = 1. \tag{4.3.2}$$

■

By restricting ourselves to nonempty connected maps—that is, to maps with exactly one component—from the above formula we obtain the famous *Euler polyhedral formula*, which is the classical formula for the *Euler characteristic* of a 2-dimensional sphere with a cell decomposition.

Euler Polyhedral Formula: If $\mathcal{L}$ is a nonempty connected map, then

$$v_{\mathcal{L}} - e_{\mathcal{L}} + f_{\mathcal{L}} = 2. \tag{4.3.3}$$

The name given to this equation requires some clarification. First of all, why call it a "polyhedral formula" when maps are objects in the plane, whereas polyhedra are spatial objects? In general, of course, a *polyhedron* is understood to be a 3-dimensional object whose boundary consists of plane surfaces having polygons for their boundaries. The bounding surfaces are called the *faces* of the polyhedron. The boundary of a face consists of the *edges* and the *vertices* of the polyhedron. The Euler polyhedral formula holds for polyhedra that are homeomorphic to a sphere. The boundary of one such polyhedron can be identified, therefore, with a map on the sphere, that, again with the help of the stereographic projection, can be transformed into a plane map. The best-known examples to which the formula applies are the five *regular solids*. The numbers of their vertices, edges, and faces are contained in the following table:

Type	v	e	f
Tetrahedron (three-sided pyramid)	4	6	4
Hexahedron (cube)	8	12	6
Octahedron	6	12	8
Dodecahedron	20	30	12
Icosahedron	12	30	20

In addition, one would naturally like to know how close a connection to Euler it really has. On that topic, much has already been written. What is certain is that Euler knew about the formula. The first time that he mentioned it was in a letter to Goldbach [EULER 1750]. It was done, however, with the proviso, *"I cannot yet thoroughly rigorously demonstrate the following proposition."* Nevertheless, in the meeting of the Petersburg Academy on September 9, 1751, Euler presented a thoroughly complicated proof of it for convex polyhedra [EULER 1758]. That the formula had already been discovered by Descartes is pure fantasy. He could not possibly have formulated it because he had never once referred to the concept of an *edge* [FEDERICO 1982].

The accrediting of the Euler polyhedral formula to the arithmetician Faulhaber, from the city of Ulm, who supposedly communicated knowledge of it to Descartes [WUSSING and ARNOLD 1978, page 168], is even more in the realm of legend. It is highly doubtful whether Faulhaber and Descartes had ever even met [SCHNEIDER 1991, 1993].

4.4 Duality

At the end of his first "proof," Kempe remarked that one could formulate the coloring problem of maps in terms of vertices and edges of (plane) graphs only [KEMPE 1879a, page 200]. He used the ancient notion of "duality" of regular solids that had already been discussed in the so-called 15th Book of Euclid ([SCHREIBER 1987, page 74]). It is also conceivable that Kempe was familiar with Maxwell's works in which voltage ratios in electrical circuits were analyzed using the notion of duality [MAXWELL 1864, 1869]. He, however, did not pursue this line of thought for the Four-Color Problem any further than that. Instead, he used these ideas in his general graph-theoretical research program in which he wanted to study "admissible vertex colorings" of arbitrary graphs with n colors. Tait also played with the notion of duality—however, without any success [TAIT 1880, page 502]. It was Heffter [1891] who made the first real use of dual graphs. Heesch, too, had fully taken note of and exploited the process of dualization [HEESCH 1969, page 16]. In this context as well it is advisable to prepare oneself for the formal construction by an intuitive analogy. In each country, one chooses a capital city. Then one links the capital cities of neighboring countries with railway lines that cannot traverse each other and that pass through precisely one border crossing. Capital cities and railway lines form the vertices and edges of a new graph, namely, a "dual" graph for the map.

Formally, we now proceed as follows. Let a map $\mathcal{L}$ be given. Denote by L_r, $r \in \{1, 2, \ldots, f_{\mathcal{L}}\}$, the countries of $\mathcal{L}$ and by $\tilde{k}$ the number of pairs of neighboring countries. In each country L_r we choose a point $\mathbf{x}_r$. In addition, for each pair of neighboring countries, we pick a common borderline B_s and a point $\mathbf{y}_s \in \overset{\circ}{B}_s$, $s \in \{1, 2, \ldots, \tilde{k}\}$. Finally, we choose, for each country L_r and each point $\mathbf{y}_s$ in its boundary, an arc B_{rs} that joins $\mathbf{x}_r$ to $\mathbf{y}_s$ and that lies entirely in L_r, except for the end point $\mathbf{y}_s$. These arcs are chosen in such a way that no two of them have an interior point in common (Proposition 2.7.1). For each pair L_{r_1}, L_{r_2} of neighboring countries, we then obtain, by combining two adjacent arcs, the arc

$$B^*_{r_1 r_2} = B_{r_1 s} \cup B_{r_2 s}$$

that joins $\mathbf{x}_{r_1}$ with $\mathbf{x}_{r_2}$. The set of the thus constructed arcs we denote by $\mathcal{L}^*$.

Lemma 4.4.1

$\mathcal{L}^*$ is a connected map.   ■

Proof Because a common borderline belongs to the borders of exactly two countries, every point $\mathbf{y}_s$ is a border point of exactly two countries L_{r_1}, L_{r_2} and hence an interior point of exactly one arc $B^*_{r_1 r_2}$. From the way the arcs B_{rs} were chosen, it follows immediately that no two distinct arcs $B^*_{r_1 r_2}$ have an interior point in common. By the construction, only one common border point $\mathbf{y}_s$ was chosen for each pair of neighboring countries. Hence, two distinct arcs $B^*_{r_1 r_2}$ also have only one end point in common. Therefore, $\mathcal{L}^*$ is a map.

Now only the connectedness of $\mathcal{L}^*$ remains to be proved. If $\mathcal{L}$ has only one country, then $\mathcal{L}^*$ is empty and therefore connected. Otherwise, every country of $\mathcal{L}$ has a neighbor (Lemma 2.6.11). Therefore, each of the points $\mathbf{x}_r$ is a vertex of $\mathcal{L}^*$. We must show that any two of these points that are distinct from each other can be joined by an arc consisting of concatenated edges of $\mathcal{L}^*$ (Definition 2.3.6). For that, let $\mathbf{x}_{r_1}$ and $\mathbf{x}_{r_2}$ (with $r_1 \neq r_2$) be given. We choose an arc B that joins $\mathbf{x}_{r_1}$ and $\mathbf{x}_{r_2}$ but that contains no vertex of $\mathcal{L}$. Basically, we travel on a straight line from $\mathbf{x}_{r_1}$ to $\mathbf{x}_{r_2}$, except that we make detours around vertices that we may encounter by traveling along semicircular arcs centered at them. In this fashion, when we leave one country, we always step into a neighboring country. These constructed edges of $\mathcal{L}^*$ from one neighbor to another concatenate to form the desired arc.　　□

The map $\mathcal{L}^*$ obtained in this way from a fixed map $\mathcal{L}$ is said to be *dual to* $\mathcal{L}$. We emphasize that this construction is in no way unique. In fact, there exist many maps that are dual to a given fixed map.

The notion of duality between regular solids, mentioned at the beginning of this section, fits in here. In order to recognize this, it is better to consider the situation in terms of the sphere. We take a regular solid. It has a center and a circumscribing sphere. Then we project the edges of the solid radially from the center onto the surrounding sphere. We thus obtain a map on the spherical surface. Geometrically speaking, every country on the sphere determined in this fashion has a true center. Now we link the centers of two neighboring countries with a suitably large circular arc, and we obtain the dual map. If instead of this we join the corresponding centers with line segments, we obtain the edgewise skeletal structure of a regular solid. By projecting radially from the center, this yields the dual map. A repetition of the process yields the first regular solid back again. Two regular solids that are related to one another in this way are said to be *dual to one another*. For instance, the regular tetrahedron is dual to itself. The cube and the octahedron are dual to each other. Likewise, the dodecahedron and the icosahedron are dual to each other. The latter fact perhaps explains why the "icosian game" (see page 136) is played on the map of a dodecahedron.

In mathematical terminology, the word "dual" generally suggests a type of reversible interrelationship of the kind we have just described for regular solids. One expects something to the effect that from the formation of a dual map to an already dual map, one retrieves the original map back again. In other words, the maps $\mathcal{L}$ and $\mathcal{L}^*$ should have a kind of reciprocal relationship to one another. This cannot be true, in general, since a dual map—as already observed—is always connected whether the original map is connected or not. The use of the word "dual" is, nonetheless, justified because under certain—not too specialized—conditions, such a there-and-back movement is still possible. This has implications for the Four-Color Theorem, since the coloring problem of complete regular maps turns out to be, as is described in the next section, equivalent to a certain coloring problem of saturated graphs. The latter can, in general, be represented with more clarity than the former.

Before we implement these thoughts in greater detail, we shall carry the idea of dual maps to a still somewhat more formal level.

Definition 4.4.2
The map $\mathcal{L}^*$ is said to be *dual* to the map $\mathcal{L}$ if the following conditions hold:
1. No vertex of $\mathcal{L}^*$ is a neutral point of $\mathcal{L}$.
2. Each country of $\mathcal{L}$ contains exactly one vertex of $\mathcal{L}^*$.
3. Two vertices of $\mathcal{L}^*$ are joined by an edge in $\mathcal{L}^*$ if and only if they lie in neighboring countries of $\mathcal{L}$.
4. An edge of $\mathcal{L}^*$ contains only points of the two countries of $\mathcal{L}$ to which its vertices belong and contains exactly one interior point of a common borderline of these countries.

Lemma 4.4.1 ensures the existence of dual maps, in the sense of this definition, for maps with at least two countries. Over and above that, it is easy to see that the empty map is dual to the maps with exactly one country. We now note that:

Lemma 4.4.3
If the map $\mathcal{L}^*$ is dual to the map $\mathcal{L}$, then an edge of $\mathcal{L}$ contains at most one point of the neutrality set of $\mathcal{L}^*$. ∎

Proof If the edge $B \in \mathcal{L}$ were to intersect two edges $B_1^*, B_2^* \in \mathcal{L}^*$, then B_1^* and B_2^* would have the same end points. This is not possible because multiple edges are not allowed. □

In considering dual maps topologically, nothing really terrible can go wrong.

Lemma 4.4.4
Let $\mathcal{L}$ be a map and let $\mathcal{L}^*$ be a map dual to $\mathcal{L}$. Then there exists a map $\widetilde{\mathcal{L}}$ such that

$$N_{\widetilde{\mathcal{L}}} = N_{\mathcal{L}} \cup N_{\mathcal{L}^*}.$$

∎

Proof We must specify which edges belong to the map $\widetilde{\mathcal{L}}$. As edges of $\widetilde{\mathcal{L}}$, we include:

1. the edges in $\mathcal{L}$ that have no point in common with $N_{\mathcal{L}^*}$;

2. the subarcs into which the edges of $\mathcal{L}$ have been subdivided by an intersection point with an edge in $\mathcal{L}^*$1

3. the subarcs into which the edges of $\mathcal{L}^*$ have been subdivided by an intersection point with an edge of $\mathcal{L}$.

It is immediately clear that these sets of edges together form a map with the desired property.   □

The impact of this lemma lies in the fact that, from the simultaneous consideration of two maps that are dual to one another, we can always proceed on the assumption that both maps consist of polygonal arcs (Theorem 2.3.9). From the Wagner and Fáry theorem (Theorem 4.2.11), we can even assume that the given map $\widetilde{\mathcal{L}}$ consists only of line segments—meaning that the polygonal arcs of the maps $\mathcal{L}$ and $\mathcal{L}^*$ each comprise at most two adjacent line segments. This is what we want to assume in what is to follow.

An application of this special case implies the following:

Lemma 4.4.5
If $\mathcal{L}$ is a nonempty map and $\mathcal{L}^*$ is a map dual to $\mathcal{L}$, then every country of $\mathcal{L}^*$ contains (at least) one vertex of $\mathcal{L}$.   ∎

Proof Let $\mathcal{L}^*$ be the empty map. Then the single country in $\mathcal{L}^*$ contains all vertices of the nonempty, by assumption, map $\mathcal{L}$.

Now let $\mathcal{L}^*$ be nonempty, and let L^* be a country of $\mathcal{L}^*$. We choose a borderline B^* of L^*. Denote by B the unique specified edge of $\mathcal{L}$ that intersects B^*. Call the intersection point $\mathbf{y}$. By the previous lemma, we can assume that $\mathbf{y}$ subdivides the edges B and B^* into *line segments* $[\mathbf{x}_1, \mathbf{y}]$ and $[\mathbf{x}_2, \mathbf{y}]$, respectively $[\mathbf{x}_1^*, \mathbf{y}]$ and $[\mathbf{x}_2^*, \mathbf{y}]$. We choose a closed disk D centered at $\mathbf{y}$ that has only interior points of the edges B and B^* in common with the union of the neutrality sets $N_{\mathcal{L}} \cup N_{\mathcal{L}^*}$. Denote by $\mathbf{z}_1, \mathbf{z}_2, \mathbf{z}_1^*, \mathbf{z}_2^*$ the intersection points of the line segments mentioned above (in the specified order) with the circumference K of D. Because $\mathbf{z}_1^*$ and $\mathbf{z}_2^*$ lie in distinct countries of $\mathcal{L}$, the pairs $\{\mathbf{z}_1, \mathbf{z}_2\}$ and $\{\mathbf{z}_1^*, \mathbf{z}_2^*\}$ are separated in K. The circle K is subdivided by $\mathbf{z}_1^*$ and $\mathbf{z}_2^*$ into two arcs K_1 and K_2, of which at least one, say K_1, lies in L^*.

In view of the separation property, we can assume that $\mathbf{z}_1 \in K_1$ and $\mathbf{z}_2 \in K_2$ and hence that $\mathbf{z}_1 \in L^*$. Since the line segment $[\mathbf{x}_1, \mathbf{z}_1]$ has no point in common with the neutrality set $N_{\mathcal{L}^*}$ (Lemma 4.4.3), it follows that L^* contains the vertex $\mathbf{x}_1$ of $\mathcal{L}$. □

At this point, we count the vertices, edges, and countries of maps dual to one another.

Lemma 4.4.6

Let $\mathcal{L}$ be a map with at least two countries and $\mathcal{L}^*$ a map dual to $\mathcal{L}$. Then:

$$v_{\mathcal{L}^*} = f_{\mathcal{L}}, \tag{4.4.1}$$

$$e_{\mathcal{L}^*} \le e_{\mathcal{L}}, \tag{4.4.2}$$

$$f_{\mathcal{L}^*} \le v_{\mathcal{L}}. \tag{4.4.3}$$

If $\mathcal{L}$ is regular, then the inequalities 4.4.2 and 4.4.3 become identities. ■

Proof Identity (4.4.1) follows directly from condition 2 of the definition of dual maps (Definition 4.4.2).

For each pair of neighboring countries of $\mathcal{L}$, we have exactly one edge in $\mathcal{L}^*$. Each such pair has at least one common borderline. Moreover, there can still be bridges and final edges in $\mathcal{L}$. Hence, inequality (4.4.2) is true. In a regular map (Definition 3.2.8), every pair of neighboring countries has exactly one common borderline; bridges and final edges do not crop up. This implies equality of the numbers of edges in the case of regular maps.

Inequality (4.4.3) follows from the fact that each country of $\mathcal{L}^*$ contains at least one vertex of $\mathcal{L}$ (Lemma 4.4.5). Since a regular map $\mathcal{L}$ is connected, it follows that $z_{\mathcal{L}} = 1$, and inequality (4.4.2) becomes an equality. For such a map $\mathcal{L}$, essentially because of the Euler polyhedral formula (4.3.3), the following holds:

$$f_{\mathcal{L}^*} = 2 + e_{\mathcal{L}^*} - v_{\mathcal{L}^*} = 2 + e_{\mathcal{L}} - f_{\mathcal{L}} = v_{\mathcal{L}}. \qquad □$$

These counting arguments, which are basically quite simple, have important consequences.

Theorem 4.4.7

Let $\mathcal{L}$ be a regular map and $\mathcal{L}^*$ a map dual to $\mathcal{L}$.

(a) *The map $\mathcal{L}$ is a map dual to $\mathcal{L}^*$.*
(b) *The map $\mathcal{L}^*$ is regular.* ∎

Proof Since $\mathcal{L}$, by assumption, is regular, every edge $B \in \mathcal{L}$ is always a unique common borderline of two countries of $\mathcal{L}$. Therefore, for each edge B, there exists an edge $B^* \in \mathcal{L}^*$ that intersects B in a single point. The point is an interior point not only of B but also of B^*. From an abstract point of view, this yields a bijection $\mathcal{L} \to \mathcal{L}^*$ in which the image of the edge $B \in \mathcal{L}$ is denoted, in general, by B^*. In what follows, an arbitrary edge of $\mathcal{L}^*$ will be denoted by a symbol representing the edge with $*$ as a superscript, which, when the $*$ is removed, will denote the corresponding edge of $\mathcal{L}$.

(a) One must show that $\mathcal{L}$ satisfies the conditions of Definition 4.4.2. A vertex of $\mathcal{L}$ is not a neutral point of $\mathcal{L}^*$ because the edges in $\mathcal{L}^*$ contain only points of countries and interior points of edges in $\mathcal{L}$ (condition 4 of Definition 4.4.2).

Since a vertex of $\mathcal{L}$ lies in each country of $\mathcal{L}^*$ (Lemma 4.4.5) and $\mathcal{L}^*$ has exactly as many countries as $\mathcal{L}$ has vertices (Lemma 4.4.6), then only one vertex of $\mathcal{L}$ can lie in each country of $\mathcal{L}^*$. Thus condition 2 of Definition 4.4.2 is satisfied.

Next, we must show that two vertices in $\mathcal{L}$ are joined by an edge in $\mathcal{L}$ if and only if they lie in neighboring countries of $\mathcal{L}^*$. Let two vertices $\mathbf{x}_1$, $\mathbf{x}_2$ be given. On the basis of the previous proofs, $\mathbf{x}_1$ and $\mathbf{x}_2$ lie in distinct countries L_1^*, respectively L_2^*, of $\mathcal{L}^*$. If B is an edge in $\mathcal{L}$ that joins $\mathbf{x}_1$ and $\mathbf{x}_2$, then B^* is a common borderline of L_1^* and L_2^*. Hence, the two countries are neighbors. Conversely, if B^* is a common borderline of L_1^* and L_2^*, then the end points of B lie in distinct countries of $\mathcal{L}^*$ whose borders contain B^*. Therefore, the points $\mathbf{x}_1$ and $\mathbf{x}_2$ must be the end points of B, and hence they are joined by an edge in $\mathcal{L}$.

From this it is also clear that an edge of $\mathcal{L}$ contains only points in those countries of $\mathcal{L}^*$ in which their end points lie. Thus each edge contains exactly one point on the common borderline.

(b) The conditions of regularity (Definition 3.2.8) must be shown for $\mathcal{L}^*$. Because $\mathcal{L}$ has at least four countries (Lemma 3.2.11), there exist countries which are neighbors of one another (Lemma 2.6.11).

Hence, $\mathcal{L}^*$ is not empty (condition 3 of Definition 4.4.2). We have already shown the connectedness of $\mathcal{L}^*$ (Lemma 4.4.1).

From the proof of (a), it follows that each edge of $\mathcal{L}^*$ is a common borderline of two different countries. Therefore, it is a circuit edge (Corollary 2.6.7). Finally, if B^* is a common borderline of two countries, then B joins vertices of $\mathcal{L}$ that lie in these countries. Since two vertices of $\mathcal{L}$ are joined by at most one edge in $\mathcal{L}$, two countries of $\mathcal{L}^*$ have at most one common borderline. □

Remark: A somewhat more elegant theory of duality emerges if one allows multiple edges and loops. Then, in the construction of a dual graph, one chooses a "dual" edge for *each* edge—for bridges and final edges, even for loops. Countries having several common borderlines generate multiple edges. This also holds conversely. In dual graphs, bridges and final edges develop out of loops. Multiple edges give rise to countries with several common borderlines. In the context of such a theory, a graph is then already dual to each of its dual graphs if it is nonempty and connected (compare part (a) of Theorem 4.4.7). In this case, the property of regularity is no longer necessary.

4.5 Cubic Maps

In this section we want to close in on the minimal criminals even further. We begin with an important, but sometimes, in respect to its import for the Four-Color Theorem, overrated, application to the construction of dual maps. The history behind this theorem and its import have already been completely detailed in Chapter 1 (see page 21).

Theorem 4.5.1 (Weiske Theorem)
There exists no map with five pairwise neighboring countries. ■

Proof Let $\mathcal{L}$ be a map that contains five pairwise neighboring countries. Then a map $\mathcal{L}^*$ dual to $\mathcal{L}$ contains five vertices that are pairwise joined by edges—in other words, a complete graph with five vertices. However, such a graph does not exist (Theorem 4.1.2)! □

In regard to the Four-Color Theorem, the following result emerges immediately.

Corollary 4.5.2

A minimal criminal has at least six countries.   ■

Proof Suppose we are given a map with five countries. We then color two nonneighboring countries with color 1. The remaining three countries require only three extra colors.   □

The Weiske theorem takes on a deeper significance through the next assertion.

Corollary 4.5.3

If a country of an arbitrary map has more than three neighbors, then it has two neighbors that have no common borderline.   ■

Proof Suppose that in the map $\mathcal{L}$, the country L_0 has at least the following pairwise distinct neighbors, $L_1, \ldots, L_4$. The Weiske theorem implies that out of the five countries $L_0, \ldots, L_4$, two cannot be neighbors of each other. From the assumption, neither of these countries can be L_0.   □

We take note of two interesting consequences of this result in relation to our hunt for criminals.

Theorem 4.5.4

In a minimal criminal, no countries have fewer than five distinct neighbors.   ■

Proof Let $\mathcal{L}$ be a minimal criminal. We already know that there cannot exist countries in $\mathcal{L}$ with fewer than four neighbors (Proposition 3.2.1). One possibility remains: that there exists a country L_0 with precisely four distinct neighbors $L_1, \ldots, L_4$. We can assume that L_1 and L_3 have no common borderline (Corollary 4.5.3). By removing the common borderlines of L_0 with L_1 and L_3, we obtain a map $\mathcal{L}'$ in which the countries L_0, L_1, and L_3 are merged into one country L'. In this construction, the remaining countries are not altered. Since $\mathcal{L}'$ has two countries fewer than $\mathcal{L}$, $\mathcal{L}'$ has an admissible 4-coloring. From this we can obtain an admissible 4-coloring for $\mathcal{L}$ as follows. First, we assign L_1 and L_3 the color given to L'. The remaining countries that are distinct from L_0 retain the color assigned to them by the coloring of $\mathcal{L}'$. Hence, for the neighbors of L_0, only three colors are needed. The fourth color remains at our disposal for the coloring of L_0. Consequently, $\mathcal{L}$ cannot be a criminal. $\qquad\square$

The next result is of a foundational nature.

Theorem 4.5.5
Every vertex of a minimal criminal has degree 3. $\qquad\blacksquare$

Proof Every vertex of a regular map has at least degree 3 (Lemma 3.2.10). What remains to be shown is that the degree of any vertex in a minimal criminal can be at most 3.

Let $\mathcal{L}$ be a minimal criminal. Assume that there exists a vertex in $\mathcal{L}$ such that $d_{\mathcal{L}}(\mathbf{x}) > 3$. We choose an elementary neighborhood D of $\mathbf{x}$. Then we construct a new map $\mathcal{L}'$ in which we include the following as new edges: firstly, those edges in $\mathcal{L}$ having $\mathbf{x}$ as end point that have been shortened by cutting off the portion of their line segments projecting into D and, secondly, the circular arcs into which the circumference of D is subdivided by these edges. In doing this, each country with $\mathbf{x}$ as a border point will be reduced in size by a sector of D. The interior L_0' of D is then included as a new country.

Now the number of countries with $\mathbf{x}$ as border point is equal to the degree of the vertex (Corollary 3.2.13). By the construction, this is the number of neighbors of L_0'. Since $d_{\mathcal{L}}(\mathbf{x}) > 3$, we can find neighbors L_1' and L_2' of L_0' that have no common borderline (Corollary 4.5.3). We note that the countries L_1 and L_2, from which L_1' and L_2' have been derived in the transition from $\mathcal{L}$ to $\mathcal{L}'$ through this reduction process, also have no common borderline. They simply abut at the vertex $\mathbf{x}$.

Now we proceed in a similar fashion to the previous proof. By removing the common borderlines of L_0' with L_1' and L_2', we obtain a map $\mathcal{L}''$ in which the countries L_0', L_1', and L_2' are merged into a single country L''. The remaining countries are not changed in the transition from $\mathcal{L}'$ to $\mathcal{L}''$. Since $\mathcal{L}''$ always has one country fewer than $\mathcal{L}'$, $\mathcal{L}''$ has an admissible 4-coloring φ''. From this we can obtain an admissible 4-coloring of $\mathcal{L}$. We assign the color of L'' to the countries L_1 and L_2. All other countries are assigned the colors of the portions of their countries lying outside of D according to the coloring φ''. Consequently, $\mathcal{L}$ cannot be a minimal criminal. This contradicts the given assumption. Hence, no such vertex exists. $\qquad\square$

Remark: Cayley already had had the idea of "inflating" vertices at which too many edges abut [CAYLEY 1879]. Kempe, too, had used this technique [KEMPE 1879a]. However, both had developed the technique only up to the construction of $\mathcal{L}'$. In doing so, they increased the number of countries. Story pointed out that a "neutral cost" method existed [STORY 1879]. With this method, he merged L_0' with one of the countries of $\mathcal{L}'$ bordering on L_0'. The map thus constructed would naturally also be a minimal counterexample if the original map $\mathcal{L}$ had been one, and then one could work further with it. Our result, which was first completely formulated by Birkhoff [BIRKHOFF 1913], is more powerful, since it implies that the map $\mathcal{L}$ cannot in any way be a minimal criminal.

Theorem 4.5.5 motivates the following concept.

Definition 4.5.6
A map is said to be *cubic* if it is regular and if all of its vertices have degree precisely equal to 3.

Theorem 4.5.5 implies that any minimal counterexample is cubic. We make note of the following properties of cubic maps in relation to the notion of duality.

Theorem 4.5.7
Let $\mathcal{L}$ be a regular map and $\mathcal{L}^*$ a map dual to $\mathcal{L}$.
(a) *$\mathcal{L}^*$ is saturated if and only if $\mathcal{L}$ is cubic.*
(b) *$\mathcal{L}^*$ is cubic if and only if $\mathcal{L}$ is saturated.*   ∎

Proof (a) Let $\mathcal{L}$ be cubic. We must show that the national borders of $\mathcal{L}^*$ are triangles (Theorem 4.2.4). We consider a country L^* of $\mathcal{L}^*$. We denote by $\mathbf{x}$ the vertex of $\mathcal{L}$ lying in L^*. It is the end point of exactly three edges B_1, B_2, and B_3 in $\mathcal{L}$. The edges B_1^*, B_2^*, and B_3^* form a circuit in $\mathcal{L}^*$, and hence they constitute the national border of L^*.

Conversely, let $\mathcal{L}^*$ be saturated. Then it must be shown that every vertex of $\mathcal{L}$ has degree 3. We consider a vertex $\mathbf{x}$ of $\mathcal{L}$ and denote by L^* the country of $\mathcal{L}^*$ in which $\mathbf{x}$ lies. The border of L^* consists of exactly three edges in $\mathcal{L}^*$. Every edge in $\mathcal{L}$ incident with $\mathbf{x}$ must intersect an edge in $\mathcal{G}_{L^*}$. However, every edge in $\mathcal{G}_{L^*}$ will be intersected by at most one edge in $\mathcal{L}$. Therefore, $d_{\mathcal{L}}(\mathbf{x}) \leq 3$. Because $\mathcal{L}$ is assumed to be regular, it follows that $d_{\mathcal{L}}(\mathbf{x}) = 3$.

(b) The proof of (b) follows from the proof of (a) because $\mathcal{L}$ is dual to $\mathcal{L}^*$ (part (a) of Theorem 4.4.7). $\qquad\square$

Note 4.5.8
On the basis of the theorem proved above, one can deduce the following equivalences:

$$\mathcal{L} \text{ cubic} \quad \Leftrightarrow \quad \mathcal{L}^* \text{ saturated,}$$
$$\mathcal{L} \text{ saturated} \quad \Leftrightarrow \quad \mathcal{L}^* \text{ cubic.}$$

However, these do not hold in general, only, as is formulated in our theorem, under the assumption of regularity of $\mathcal{L}$. Here is a counterexample.

The graph illustrated above is obviously not saturated. However, every graph dual to it is cubic. $\qquad\blacksquare$

4.6 Some Counting Arguments

One can establish basic estimates as to the numbers of vertices, edges, and countries of a map. More important, however, are a few of the relationships among these numbers. In that regard, we consider a map $\mathcal{L}$ with the vertices $\mathbf{x}_1, \mathbf{x}_2, \ldots, \mathbf{x}_{v_\mathcal{L}}$ and the countries L_1, $L_2, \ldots, L_{f_\mathcal{L}}$. In addition, we set $d_r = d_\mathcal{L}(\mathbf{x}_r)$ for $r \in \{1, \ldots, v_\mathcal{L}\}$ and let n_s denote the number of borderlines of L_s for $s \in \{1, \ldots, f_\mathcal{L}\}$.

Let $\mathcal{L}$ be a regular map. Then the following hold:

$$v_\mathcal{L} \geq 4, \quad e_\mathcal{L} \geq 6, \quad f_\mathcal{L} \geq 4, \tag{4.6.1}$$

$$d_r \geq 3 \text{ for all } r, \qquad\qquad n_s \geq 3 \text{ for all } s, \tag{4.6.2}$$

$$\sum_{r=1}^{v_\mathcal{L}} d_r = 2 \cdot e_\mathcal{L}, \qquad\qquad \sum_{s=1}^{f_\mathcal{L}} n_s = 2 \cdot e_\mathcal{L}, \tag{4.6.3}$$

$$3 \cdot v_\mathcal{L} \leq 2 \cdot e_\mathcal{L}, \qquad\qquad 3 \cdot f_\mathcal{L} \leq 2 \cdot e_\mathcal{L}, \tag{4.6.4}$$

$$\mathcal{L} \text{ cubic} \Rightarrow \qquad\qquad \mathcal{L} \text{ saturated} \Rightarrow$$
$$3 \cdot v_\mathcal{L} = 2 \cdot e_\mathcal{L}, \qquad\qquad 3 \cdot f_\mathcal{L} = 2 \cdot e_\mathcal{L}, \tag{4.6.5}$$

Proof We have already shown that a regular map $\mathcal{L}$ has at least four countries (Lemma 3.2.11). Since regularity is carried over to dual maps (part (b) of Theorem 4.4.7), $\mathcal{L}^*$ is also regular. Hence, $\mathcal{L}$ has at least four vertices (inequality (4.4.3)). The number of edges can be deduced from the Euler polyhedral formula (4.3.3):

$$e_\mathcal{L} = v_\mathcal{L} + f_\mathcal{L} - 2 \geq 4 + 4 - 2 = 6 .$$

In this way, the lower bounds (4.6.1) are established.

As for inequalities and equalities (4.6.2)–(4.6.5), we first prove the results in the left column. Now we have just proven that at least three edges are incident with each vertex of a regular map (Lemma 3.2.10). This implies inequality (4.6.2 left). Each edge has two end points. Consequently, every edge is counted twice on the left side of the identity (4.6.3 left). In order to show (4.6.4 left), we first note that we can interpret each natural number v as a sum of v ones:

$$v = \underbrace{1 + 1 + \cdots + 1}_{v \text{ summands}} = \sum_{r=1}^{v} 1 .$$

Using this, we can estimate the following:

$$
\begin{aligned}
3 \cdot v_\mathcal{L} &= 3 \cdot \sum_{r=1}^{v_\mathcal{L}} 1 = \sum_{r=1}^{v_\mathcal{L}} 3 && \text{because of (4.6.2 left)} \\
&\leq \sum_{r=1}^{v_\mathcal{L}} d_r && \text{because of (4.6.3 left)} \\
&= 2 \cdot e_\mathcal{L}.
\end{aligned}
$$

If $\mathcal{L}$ is cubic, then $d_r = 3$ for all $r \in \{1, \ldots, v_\mathcal{L}\}$, and we obtain (4.6.5 left).

To establish (4.6.2 right) to (4.6.4 right), we take a map $\mathcal{L}^*$ dual to $\mathcal{L}$. We denote by $\mathbf{x}_s^*$ the vertex of $\mathcal{L}^*$ lying in L_s and by d_s^* the degree of $\mathbf{x}_s^*$ for $s \in \{1, \ldots, v_{\mathcal{L}^*} = f_\mathcal{L}\}$ (see equality (4.4.1)). By assumption, $\mathcal{L}^*$ is regular. Therefore, L_s has n_s distinct neighbors, and so $d_s^* = n_s$ for all s (condition 3 of Definition 4.4.2). Again, because of regularity, $e_{\mathcal{L}^*} = e_\mathcal{L}$ and $f_{\mathcal{L}^*} = v_\mathcal{L}$ (Lemma 4.4.6).

We now list the (in-)equalities (4.6.2 left)–(4.6.5 left) for the map $\mathcal{L}^*$. Then we replace the starred quantities with the corresponding unstarred ones. In this way, we obtain the (in-)equalities (4.6.2 right)–(4.6.5 right).   □

It is worth taking note of the following consequence of identity (4.6.5).

Corollary 4.6.1
A cubic map has an even number of vertices. A saturated map has an even number of countries. ∎

Proof This follows from the theorem on the unique factorization of natural numbers into primes. The right sides of the equalities (4.6.5) are, respectively, even numbers. Thus the left sides must also be even numbers. □

There exist yet two more estimates that are of such fundamental importance to the Four-Color Theorem that we list them separately.

Theorem 4.6.2
The following hold for a regular map:

$$\sum_{r=1}^{v_{\mathcal{L}}}(6 - d_r) \geq 12, \qquad\qquad (4.6.6)$$

$$\sum_{s=1}^{f_{\mathcal{L}}}(6 - n_s) \geq 12. \qquad\qquad (4.6.7)$$
∎

Proof It suffices to verify only one of these inequalities. As has been shown in the previous proof, the other can be verified by considering a dual map. We do the following calculations:

$$
\begin{aligned}
\sum_{r=1}^{v_{\mathcal{L}}}(6 - d_r) \;&=\; 6 \cdot v_{\mathcal{L}} - 2 \cdot e_{\mathcal{L}} && \text{because of (4.6.3 left)} \\
&=\; 6 \cdot v_{\mathcal{L}} - 6 \cdot e_{\mathcal{L}} + 4 \cdot e_{\mathcal{L}} \\
&\geq\; 6 \cdot v_{\mathcal{L}} - 6 \cdot e_{\mathcal{L}} + 6 \cdot f_{\mathcal{L}} && \text{because of (4.6.4 right)} \\
&=\; 6 \cdot (v_{\mathcal{L}} - e_{\mathcal{L}} + f_{\mathcal{L}}) \\
&=\; 12 \text{ from the Euler Polyhedral Formula (4.3.3).} \quad □
\end{aligned}
$$

One can postulate that the entire proof of the Four-Color Theorem could be built around the inequalities proven above. However, this is somewhat exaggerated. Kempe, however, did use an important corollary deduced from them [KEMPE 1879a, page 198]. The right sides of the inequalities (4.6.6) and (4.6.7) are positive. Therefore, the sums on the left sides must also be positive. From this, it follows that:

Corollary 4.6.3
Every regular map has vertices that are incident with at most five edges and countries having at most five neighbors. ∎

Furthermore, there emerges an improved, in comparison to Corollary 4.5.2, lower limit for the number of countries in a minimal criminal.

Corollary 4.6.4
A minimal criminal has at least thirteen countries. ∎

Proof Since every country in a minimal criminal has at least five neighbors (Theorem 4.5.4), the individual summands on the left side of inequality (4.6.7) are all less than or equal to 1. Because the total sum must be at least 12 (right side of this inequality), there must be at least twelve summands.

A minimal criminal is a cubic map (Theorem 4.5.5) in which every country has at least five neighbors. We now consider a map that has exactly twelve countries and that also satisfies both of these properties. Then each summand on the left side of inequality (4.6.7) must be positive. In other words, no country can have more than five neighbors. Hence, all countries have exactly five neighbors. This implies that there are exactly 30 edges (equality (4.6.3) right) and 20 vertices (equality (4.6.5) left). Consequently, up to a homeomorphism, it must be the stereographic projection of the surface of a dodecahedron. Such a map, however, can easily be colored with four colors (see the following). □

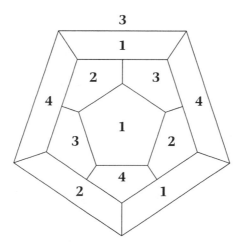

4.7 Intermezzo: The Five-Color Theorem

At this point, we will include Heawood's proof of the Five-Color Theorem.

Theorem 4.7.1 (Five-Color Theorem)
Every map has an admissible 5-coloring. ∎

Proof Here, as well, the method of the minimal counterexample (page 86) forms the basis of the proof. When considering a minimal counterexample to the Four-Color Theorem, we can exclude countries having only three neighbors (Proposition 3.2.1). In the same way, it follows that a minimal counterexample to the Five-Color Theorem can have neither countries with three neighbors nor countries with four neighbors. Moreover, it is possible, in this case as well, to restrict oneself to regular maps.

In any minimal counterexample one can find a country L_0 with five pairwise distinct neighbors $L_1, \ldots, L_5$ (Corollary 4.6.3). Among the neighbors of L_0, there are two, let's say L_2 and L_4, that have no common borderline. By removing the borderlines between L_0 and L_2 and between L_0 and L_4, we obtain a map $\mathcal{L}'$ for which we can find a 5-coloring. In the transition from $\mathcal{L}$ to $\mathcal{L}'$ we have merged the countries L_0, L_2, and L_4 into a single country L'. All other countries have remained the same. We now color $\mathcal{L}$ as follows. The countries distinct from L_0, L_2, and L_4 will be colored with the colors provided to them by the 5-coloring of $\mathcal{L}'$. The countries L_2 and L_4 will take on the color of L'. Thus only four colors are needed for the neighbors of L_0. The fifth color remains at our disposal to be applied to L_0.   □

4.8   Tait's Reformulation

We end this chapter with Tait's ideas in regard to the Four-Color Problem, which he felt constituted an alternative proof to the Four-Color Theorem. In reality, however, they provided only an equivalent version of the problem. Instead of colorings of countries, Tait considered colorings of edges.

Definition 4.8.1
Let $\mathcal{L}$ be a map and $n \in \mathbb{N}$. An *n-coloring of edges*, or an *edge n-coloring*, of $\mathcal{L}$ is a map $\psi : \mathcal{L} \to \{1, \ldots, n\}$. Such a coloring is *admissible* if edges with common end points always have distinct function values ("colors").

Tait showed that:

Theorem 4.8.2
A cubic map has an admissible 4-coloring if and only if it has an admissible 3-coloring of its edges.   ■

Proof   Let $\mathcal{L}$ be a cubic map. If a 4-coloring of $\mathcal{L}$ is given, then we can easily achieve a 3-coloring of its edges. To that end, we consider an edge $B \in \mathcal{L}$. Since $\mathcal{L}$ contains only circuit edges, B is a common borderline of exactly two countries (Lemma 2.6.2). Now we distinguish between two separate cases:
1. If one of the two bordering countries is colored with color 4, then B is colored with the color of the other bordering country.
2. If each of the bordering countries is colored with one of the colors 1, 2, or 3, then B is colored with the remaining one of these three colors that is not used for the two bordering countries.

Conversely, the construction of a 4-coloring from a 3-coloring of edges is somewhat more difficult. An elegant procedure for this, called the *Klein 4-group* [AIGNER 1984, page 22], uses algebraic methods. However, in this exposition we do not want to go into this particular method any further. From the point of view of present-day demands for precise mathematical proofs, one can only treat Tait's arguments [TAIT 1884] as *sketches of a proof*. We will present only the basic ideas that were given by Errera [1927].

Let $\mathcal{L}$ be a cubic map. Suppose $\mathcal{L}$ is given a 3-coloring of its edges. We denote by $\mathcal{L}_1$ and $\mathcal{L}_2$ the maps that we obtain from $\mathcal{L}$ by removing the edges colored with color number 1, respectively color number 2. In these new maps, every vertex has degree 2, and their neutrality sets consist, therefore, of pairwise disjoint closed Jordan curves.

We now consider the map $\mathcal{L}_1$. We give each country a label of either the letter a or the letter b. The label a is given to the unbounded country. All neighbors of the unbounded country receive the label b. In doing so, we observe that no two neighboring countries of the unbounded country are neighbors of each other. The next step is to assign the label a to the bounded countries that are neighbors of the already b-marked countries. Those still unlabeled countries that are neighbors of the countries just marked with the letter a are given the label b, and so on. Thus we achieve a labeling of the countries of $\mathcal{L}_1$ in which neighboring countries have different labels. In the same way, we label the countries of $\mathcal{L}_2$ with either a or b markings such that neighboring countries are differently labeled.

Every country L of $\mathcal{L}$ is the intersection of a country L_1 of $\mathcal{L}_1$ and L_2 of $\mathcal{L}_2$. We give the countries L of $\mathcal{L}$ labels consisting of ordered pairs (x, y) where $x, y \in \{a, b\}$ in such a way that x is the label of L_1 and y is the label given to L_2. We thus have four designations (a, a), (a, b), (b, a), and (b, b). For two neighboring countries L and L' of $\mathcal{L}$, the following cases are possible:

- The common borderline has the color 1. Then the labels in the first components of the pairs are the same, but the labels in the second components differ.
- The common borderline has color 2. Then the labels in the second components agree, but the labels in the first components are different.

- The common borderline has color 3. Then in both component slots the labels assigned to the two countries are distinct from one another.

In this way, neighboring countries always have distinct labels. We now color the (a, a) countries with the color 1, the (a, b) countries with the color 2, the (b, a) countries with the color 3, and the (b, b) countries with the color 4. We have just constructed the desired admissible 4-coloring of $\mathcal{L}$. □

Tait had also specified a situation in which a 3-coloring of edges is easy to produce.

Definition 4.8.3

A circuit $\mathcal{K}$ in a map $\mathcal{L}$ is said to be *Hamiltonian* if it traverses all vertices of $\mathcal{L}$, that is, if each vertex of $\mathcal{L}$ is an end point of (two) edges of $\mathcal{K}$.

The above notion is named in honor of Hamilton,[4] who was one of the first to study such circuits. It was even done in connection with his "icosian game" (*icosian* from the Greek word εἴκοσι, which in English means *twenty*). This game, sometimes called "around the world," consists in the task of specifying a Hamiltonian circuit along the edges of a dodecahedron that contains twenty vertices [HAMILTON 1857].

Theorem 4.8.4

If there exists a Hamiltonian circuit in a cubic map, then the map has a 3-coloring of its edges. ∎

Proof   Let $\mathcal{L}$ be a cubic map. The key to the claim to be proved is the fact that a cubic map has an even number of vertices (Corollary 4.6.1).

In a circuit, the number of edges is equal to the number of vertices. Therefore, a Hamiltonian circuit has an even number of edges, which we can color alternately with two colors. We then color the remaining edges with the third color. Thus we have an admissible 3-coloring of edges of $\mathcal{L}$, since at each vertex, two edges of the Hamiltonian circuit and a third distinct edge abut. □

---

[4]This is not completely justified, see [BIGGS–LLOYD–WILSON 1976, pages 28 ff].

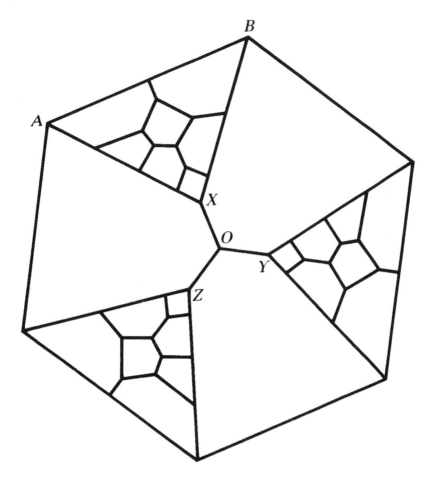

FIGURE 4.1 Example of a cubic map without a Hamiltonian circuit.
It has 69 edges, 46 vertices, and 25 countries. Tutte generated it in 1946.

Tait had conjectured that each cubic map contains a Hamiltonian
circuit. The Tait conjecture is, however, false, as is shown by the
counterexample (above) provided by Tutte [Tutte 1946]; [Aigner
1984, page 66].

5 CHAPTER

The Combinatorial Version of the Four-Color Theorem

A purely combinatorial formulation of the Four-Color Theorem will be presented in this chapter (Theorem 5.3.1). For that we still require a few preparatory ideas.

5.1 Colorings of Vertices

From a graph-theoretical standpoint, the structure of a graph is determined solely by its vertices and its edges. On the other hand, its countries—that is, its faces—are objects derived from it. It makes sense, therefore, to consider vertex colorings of a graph in conjunction with edge colorings, which were introduced at the end of the previous chapter (Definition 4.8.1).

Definition 5.1.1
Let $G = (E, \mathcal{L})$ be a graph and $n \in \mathbb{N}$. A *vertex n-coloring* of G is a mapping $\chi : E \to \{1, \ldots, n\}$.[1] It is said to be *admissible* if two vertices

---

[1]In the context of the Four-Color Theorem, for explicit computations one often uses the numbers 0, 1, 2, 3 as colors instead of 1, 2, 3, 4.

that are the end points of one and the same edge of $\mathcal{L}$ always have distinct function values (that is, "colors").

We note that isolated vertices play no role in the existence of admissible vertex n-colorings. Therefore, we can exclude such vertices in future deliberations.

There is a close connection between vertex colorings and the Four-Color Theorem.

Theorem 5.1.2
A map has an admissible 4-coloring if and only if each map dual to it has an admissible vertex 4-coloring. ∎

Proof   This can be shown directly from the definition of dual maps (Definition 4.4.2). □

The above result can be sharpened even further.

Theorem 5.1.3
The topological Four-Color Theorem (Theorem 3.1.3) is valid if and only if each graph has an admissible vertex 4-coloring. ∎

Proof   The sufficiency of the condition arises directly from the previous theorem. The proof for the necessity of the condition follows again from the method of minimal counterexamples.

We assume that the topological version of the Four-Color Theorem is true. We consider a minimal criminal $G = (E, \mathcal{L})$ that does not admit a vertex 4-coloring and in which "minimal" refers to the number of vertices. We show that we can assume that the map $\mathcal{L}$ is regular and saturated. If $\mathcal{L}$ is not saturated, then we can add finitely many edges to it in order to obtain a saturated graph $G' = (E, \mathcal{L}')$. This can be done because the number of vertices remains constant. In any event, the coloring problem will become more difficult because more conditions must be fulfilled. Therefore, we can assume from the outset that G is saturated.

A minimal criminal has at least five vertices. A saturated graph with at most two faces has at most three vertices. Hence, G has more than two faces and hence is regular (Corollary 4.2.5).

Now we choose a map $\mathcal{L}^*$ that is dual to $\mathcal{L}$. By assumption, it has an admissible 4-coloring of its countries. Since $\mathcal{L}$ is regular, $\mathcal{L}$ is also dual to $\mathcal{L}^*$ (part (a) of Theorem 4.4.7). Hence, from the 4-coloring of

$\mathcal{L}^*$ we obtain a vertex 4-coloring of G (Theorem 5.1.2). This implies that G is not a criminal, and the theorem has been proved. □

When it comes to vertex colorings, only the basic graph-theoretical concepts of vertices and edges are encountered. These, however, still have a topological structure as points in the plane and as arcs. In what follows, it will be shown that one can extricate oneself as well from these topological entanglements.

5.2 Planar Graphs

In this section, we will, for the sake of clarity, speak expressly of "plane" graphs (in contrast to "combinatorial" graphs), even though we had once agreed that the word "graph" by itself would be used to signify a plane graph (Page 61).

In defining a line graph (Definition 4.2.10), one required only the vertex set that was a finite point set and the set of all pairs of vertices that were joined by an edge. The edges themselves were uniquely defined to be the line segments between their end points. This notion leads us to a purely combinatorial definition of a graph.

Definition 5.2.1
A *combinatorial graph* is a pair $G = (E, \mathcal{L})$ consisting of finite set E and a set $\mathcal{L}$ (also finite) of two-element subsets of E. As was the case for plane graphs, the elements of E are called the *vertices* of G, and the elements of $\mathcal{L}$ are considered to be the *edges* of G.

A few of the notions of topological graph theory can be carried over without difficulty to the combinatorial approach.

Definition 5.2.2
(a) A combinatorial graph $G = (E, \mathcal{L})$ is said to be *complete* (compare Definition 4.1.1) if $\mathcal{L}$ contains all two-element subsets of E.
(b) The *degree* $d_G(\mathbf{x})$ of a vertex $\mathbf{x}$ of a combinatorial graph G is the number of edges of G to which $\mathbf{x}$ belongs (compare Definition 3.2.2).
(c) The combinatorial graph $G' = (E', \mathcal{L}')$ is a *subgraph* of the combinatorial graph $G = (E, \mathcal{L})$ if $E' \subset E$ and $\mathcal{L}' \subset \mathcal{L}$. This is the

combinatorial analogue of the fact that every subset of edges of a map forms yet another map (Page 62).

(d) The combinatorial graphs $G' = (E', \mathcal{L}')$ and $G = (E, \mathcal{L})$ are said to be *isomorphic* if there exists a bijective mapping $\gamma : E' \to E$ that induces a bijection $\mathcal{L}' \to \mathcal{L}$. Isomorphisms take the place of the homeomorphisms of maps that we have already used on a number of occasions—for instance, in the reduction to polygonal arc maps (Theorem 2.3.9).

(e) Suppose $G = (E, \mathcal{L})$ is a graph and $n \in \mathbb{N}$. An *n-coloring* of vertices, or a *vertex n-coloring*, of G is a mapping $\chi : E \to \{1, \ldots, n\}$. It is said to be *admissible* if two vertices that are the end points of one and the same edge in $\mathcal{L}$ always have distinct function values (or "colors"), that is, if the following holds for all $\mathbf{x}, \mathbf{y} \in E$:

$$(\mathbf{x}, \mathbf{y}) \in \mathcal{L} \Rightarrow \chi(\mathbf{x}) \neq \chi(\mathbf{y}).$$

Each plane graph and each map determines a combinatorial graph.

Definition 5.2.3
Let $G = (E, \mathcal{L})$ be a plane graph. The combinatorial graph *underlying* G is the graph $G^\flat = (E, \mathcal{L}^\flat)$, where $\mathcal{L}^\flat$ denotes the set of the pairs of vertices of G that are joined by an edge in G.

The question remains as to whether every combinatorial graph is the underlying graph of a plane graph. In general, this is certainly not the case, since an arbitrary finite set E need not be a point set of the plane. However, we are not putting the question forward in such strict terms. Since we can consider isomorphic combinatorial graphs as being equivalent, in reality we are interested only in whether each combinatorial graph is perhaps isomorphic to the underlying graph of a plane graph. Even here, the answer is "No!" To exemplify this, we consider the combinatorial graphs K_5 and $K_{3,3}$ defined in the following way:[2]

---

[2]We leave it to the reader to find an explanation for these symbols, which are used in standard graph-theoretical terminology.

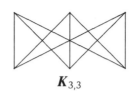

K_5 $K_{3,3}$

K_5 is the complete combinatorial graph with the numbers $1, 2, 3, 4, 5$ as its vertices.

$K_{3,3}$ has the numbers $1, 2, 3, 4, 5, 6$ as its vertices. Its edges consist of the pairs of vertices $\{g, u\}$, where g is an even number and u is odd.

Each of these combinatorial graphs, sometimes referred to as the *Kuratowski graphs,* also cannot, up to isomorphism, be the underlying graph of a plane graph. In the case of K_5, it is because there exists no complete plane graph with five vertices (Theorem 4.1.2). In the case of $K_{3,3}$, it is because the supply problem is unsolvable (Theorem 4.1.6). To see this intuitively, one needs only the representations of the Kuratowski graphs shown above. However, one must be convinced that on the basis of the appropriate theorems, a representation of these graphs that is free of overlaps is not possible—even if one allows arbitrary simple curves instead of line segments. The Wagner and Fáry theorem (Theorem 4.2.11) permits, if the occasion warrants it, a representation as a line graph. This means that one could, in the search for a better representation, limit oneself to another arrangement of the vertices in the plane. It is left to the reader to show that the combinatorial graphs that one obtains from the Kuratowski graphs by removing in each case an edge are actually isomorphic to the underlying graphs of line graphs.

The question just raised justifies the definition of another notion.

Definition 5.2.4

A *planar graph* is a combinatorial graph that is isomorphic to the underlying graph of a plane graph.

Complete combinatorial graphs having at most four vertices are planar graphs. Those with more than four vertices are not. Using this approach, the notion of being planar seems to have a topological flavor. In 1930, however, Kuratowski was able to demonstrate that

being planar was a purely combinatorial property. This will be clar-
ified in what follows. For the proof of the main theorem, however,
we refer you to the literature.

First of all, we need the concept of a subdivision of a combinatorial graph. It is reminiscent of the geometrical process of placing
border-stones (see page 60).

Definition 5.2.5
(a) The combinatorial graph $G' = (E', \mathcal{L}')$ is said to be formed from
   the combinatorial graph $G = (E, \mathcal{L})$ by a *simple subdivision* if:
   1. E' consists of E plus an additional vertex $\mathbf{z}'$, and
   2. $\mathcal{L}'$ is formed from $\mathcal{L}$ by the simultaneous removal of an edge
      $\{\mathbf{x}, \mathbf{y}\}$ and the inclusion of two new edges $\{\mathbf{x}, \mathbf{z}'\}$, $\{\mathbf{y}, \mathbf{z}'\}$.
(b) The combinatorial graph $\tilde{G}$ is said to be a *subdivision* of the
   combinatorial graph G if $\tilde{G}$ is formed from G by finitely many
   repetitions of a simple subdivision process.

We note that by the subdivision process, the degrees of the originally given vertices do not change, and the newly added vertices
always have degree 2. Hence, for example, the underlying graph
of the plane graph shown below is not isomorphic to a subdivision
of K_5.

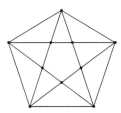

Theorem 5.2.6   (Kuratowski Theorem)
*A combinatorial graph is planar if and only if it has no subgraph that
is isomorphic to a subdivision of K_5 or $K_{3,3}$.*   ∎

The necessity of the given condition follows from the previous
discussion in this section. The sufficiency of the condition was first
proven in [KURATOWSKI 1930]. Modern variations of it and other approaches to the proof can be found in [HARARY 1969, Theorem 11.13],
[AIGNER 1984, Theorem 4.6], [WAGNER–BODENDIEK 1989, Chapter 3],
and [WAGNER–BODENDIEK 1990, Chapter 9]. Here we must dispense
with all the details. Nonetheless, we want to emphasize once again

the significance of these results. The notions of "subdivision," "isomorphism," and "subgraph" are of a purely combinatorial nature. They contain no topological entities. In order to reinforce this, we will now outline an algorithmic method to prove the planarity of a combinatorial graph. We note from the outset that this method will only illustrate how such an investigation could be done. An actual carrying out of the proof would be too cumbersome. Better algorithms are mentioned in [WAGNER–BODENDIEK 1990, Section 9.8]. First of all, however, we introduce a few notions that, in part, will take on significance only a little bit later.

Definition 5.2.7
Let $G = (E, \mathcal{L})$ be a (plane or combinatorial) graph.
(a) Two vertices of G are said to be *neighboring vertices*, or simply *neighbors*, if they are distinct from one another and are end points of the same edge in $\mathcal{L}$.
(b) A sequence $(\mathbf{x}_1, \mathbf{x}_2, \ldots, \mathbf{x}_r)$ of vertices is called a *chain (from $\mathbf{x}_1$ to $\mathbf{x}_r$)* if they are pairwise distinct and each successive pair consists of neighboring vertices. The number r is called the *length* of the chain, and the edges that join two successive vertices are called the *links* of the chain.
(c) A chain $(\mathbf{x}_1, \mathbf{x}_2, \ldots, \mathbf{x}_r)$ is said to be *simple* if two of its vertices are neighbors only if they are successive vertices in the chain. This means that for indices $j_1, j_2 \in \{1, \ldots, r\}$ where $|j_1 - j_2| > 1$, the vertices $\mathbf{x}_{j_1}$ and $\mathbf{x}_{j_2}$ are not neighbors.
(d) A chain $(\mathbf{x}_1, \mathbf{x}_2, \ldots, \mathbf{x}_r)$ is called a $d_1 - d_2 - \cdots - d_r$ *chain*, where $d_1, d_2, \ldots, d_r \in \mathbb{N}$, if for all $j \in \{1, \ldots, r\}$ the following holds:
$$d_j = d_G(\mathbf{x}_j).$$
(e) Two chains $(\mathbf{x}_1, \mathbf{x}_2, \ldots, \mathbf{x}_r), (\mathbf{x}'_1, \mathbf{x}'_2, \ldots, \mathbf{x}'_{r'})$ are said to be *disjoint* if they have no interior vertices in common. In other words, $\mathbf{x}_j \neq \mathbf{x}'_{j'}$ for all $j \in \{2, \ldots, r - 1\}$ and all $j' \in \{2, \ldots, r' - 1\}$.

Suppose we are given a combinatorial graph $G = (E, \mathcal{L})$. Then we proceed in the following way:
1. One writes down all existing chains. In doing so, we note that there are only finitely many of them. The length r of a chain is bounded by the number v of its vertices. The number of chains of a fixed length $r \in \{2, \ldots, v\}$ can be estimated as follows. There

exist at most $\binom{v}{r}$ possibilities for the set of vertices of such a chain. For any fixed set of vertices, one has at most $r!$ choices for the ordering of the chain. Hence, there exist at most

$$\binom{v}{r} \cdot r! = \frac{v!}{(v-r)!}$$

chains of length r. An upper bound for the total number of chains in G is $v! \cdot \sum_{r=2}^{v} 1/(v-r)! = v! \sum_{r=0}^{v-2} 1/r! < 3 \cdot v!$. This upper bound is as sharp as possible. A complete graph G has precisely that many chains.

2. For each choice of five vertices $x_1, \ldots, x_5$, one tries to find ten pairwise disjoint chains $C_{j_1 j_2}$ from x_{j_1} to x_{j_2}, $1 \leq j_1 < j_2 \leq 5$. Since only finitely many chains exist, one has, in this experiment, only finitely many 10-tuples of chains to examine. Therefore, one is finished in finitely many steps. If one is successful in this venture, then G contains a subdivision of K_5 and is not planar. In the event that one is not successful, the process continues.

3. For each choice of six vertices $x_1, \ldots, x_6$, one tries to find nine pairwise disjoint chains $C_{j_1 j_2}$ from x_{j_1} to x_{j_2}, $j_1 \in \{1, 3, 5\}$, $j_2 \in \{2, 4, 6\}$. If one succeeds in doing so, G contains a subdivision of $K_{3,3}$ and therefore cannot be planar. In the event that one cannot find such chains, then G is planar.

We close this section with the proof that a "contraction" of a graph does not destroy its planarity.

Definition 5.2.8
Let $G = (E, \mathcal{L})$ be a combinatorial graph. The combinatorial graph $G' = (E', \mathcal{L}')$ is formed from G:

(a) *by a simple contraction* if two vertices that are joined by an edge in G are identified with each other, that is, if there exist vertices **a** and **b** $\in E$ such that the following hold:
  1. $\{a, b\} \in \mathcal{L}$.
  2. $E' = E \setminus b$.
  3. $\mathcal{L}' = \{\{x, y\} \in \mathcal{L} : x \neq b \neq y\} \cup \{\{a, x\} \in E' : x \neq a \wedge \{x, b\} \in \mathcal{L}\}$. (Note that the individual sets of this set-theoretic union need not be disjoint!)

(b) *by contraction* if G' is obtained from G by a process of finitely many simple contractions.

Now we will show that:

Theorem 5.2.9
If the combinatorial graph $\mathbf{G'}$ is formed from a planar graph $\mathbf{G}$ by contraction, then $\mathbf{G'}$ is also planar. ∎

For the proof, we need a topological result.

Proposition 5.2.10
Let S be a line segment in $\mathbb{R}^2$. Then there exists a continuous mapping $k : \mathbb{R}^2 \to \mathbb{R}^2$ having the following properties:
1. *S is collapsed to an end point. In other words, $k(S)$ consists of a single point $\mathbf{a}$ that was one of the given end points of S.*
2. *k maps the complement of S homeomorphically onto the complement of $\mathbf{a}$. This means that k induces a homeomorphism $\mathbb{R}^2 \backslash S \longrightarrow \mathbb{R}^2 \backslash \{\mathbf{a}\}$.* ∎

Proof Without loss of generality, we can assume that S is the unit interval on the x-axis. That is,

$$S = \{(t, 0) \in \mathbb{R}^2 : t \in [0, 1]\} \, .$$

Then we can explicitly define k as follows:

$$k(x, y) = \begin{cases} (x, y) & \text{if } x \leq 0 \, , \\[2mm] \dfrac{|y|}{\sqrt{x^2 + y^2}}(x, y) & \text{if } 0 < x \leq 1 \, , \\[2mm] \dfrac{\sqrt{(x-1)^2 + y^2}}{\sqrt{x^2 + y^2}}(x, y) & \text{if } 1 \leq x \, . \end{cases}$$

It is not difficult to show that this mapping is continuous. Moreover, it is also not difficult to find the inverse mapping $\mathbb{R}^2 \setminus \{0\} \longrightarrow \mathbb{R}^2$. This we will leave to the reader. One must now only explain how the mapping k really works. It maps each line of $\mathbb{R}^2$ passing through the origin into itself in such a way that the portions of the right half-plane near the origin are contracted radially towards the origin, and the smaller the slope of the line, the greater the contraction. □

Proof of Theorem 5.2.9 It suffices to prove the claim for a simple contraction. Let $\tilde{G}$ be a plane graph whose underlying combinatorial graph is isomorphic to G. By the Wagner and Fáry theorem, we can assume that $\tilde{G}$ is a line graph. Furthermore, let S be the edge of $\tilde{G}$ whose vertices $\mathbf{a}$ and $\mathbf{b}$ (considered as vertices of G) will be identified in the transition to G'. The corollary proven above shows the existence of a continuous mapping $k : \mathbb{R}^2 \to \mathbb{R}^2$ that contracts S to $\mathbf{a}$ and that maps the complement of S homeomorphically to the plane punctured at the point $\mathbf{a}$, that is, the set $\mathbb{R}^2 \setminus \{\mathbf{a}\}$. We note that k maps all edges of $\tilde{G}$, with the exception of S, to arcs.

Now the following construction produces a plane graph $\tilde{G}'$ whose underlying graph is isomorphic to G':

- As vertices of $\tilde{G}'$, we take the images of the vertices of $\tilde{G}$ under the mapping k.
- As edges, we first choose the images under k of the edges of $\tilde{G}$ that do not have $\mathbf{b}$ as an end point.
- As edges, we finally take the images under k of the edges of $\tilde{G}$ having $\mathbf{b}$ as an end point, but whose other end point is distinct from $\mathbf{a}$ and not yet joined to $\mathbf{a}$ by an edge of $\tilde{G}$. (In this way, multiple edges in $\tilde{G}'$ are precluded.) $\qquad\square$

Remark: From another point of view, the theorem turns out to be almost a triviality. To show this, we need, however, another condition. Suppose H is a graph such that G is dual to H, that is, $G = H^*$. The theory of duality in Section 4.4 shows that this condition is not very limiting. Suppose G' is formed from G by the identification of the neighboring vertices $\mathbf{x}$ and $\mathbf{y}$. This is equivalent to the removal of a borderline between the faces in H corresponding to these vertices and hence the merger of two neighboring countries. If one were then to find the dual of the graph H' created by this merger, one would obtain a plane graph whose underlying combinatorial graph is G'. This proves that G' is planar. The translation of a simple contraction into a merger of countries provides a very helpful tool for one's intuition.

For the complete proof of the Four-Color Theorem, one really needs a polished theory of contraction (see [APPEL and HAKEN 1989, pages 178–180]). This particular topic is beyond the scope of this book and will not be discussed any further.

We now introduce yet another bit of terminology relative to this context. Practical applications of contraction occur mostly in the following way. A plane graph $\tilde{G}$, an edge B of $\tilde{G}$, and a point $\mathbf{x} \in B$

are all given. If **x** is not an end point of B, then we consider the graph that is created by subdividing B at the point **x**. We then construct the graph $\tilde{G}'$, as was done in the previous proof, by contracting either B or the two edges formed from B to **x** and by removing all resulting multiple edges. In doing so, we say that *the graph $\tilde{G}'$ is formed from $\tilde{G}$ by a geometric contraction of B at* **x**.

5.3 The Combinatorial Four-Color Theorem — Formulation and Further Advances

In the previous section, we saw that for a combinatorial graph, the property of being planar is a purely combinatorial phenomenon. From this we can now formulate the combinatorial Four-Color Theorem.

Theorem 5.3.1
Every planar graph has an admissible vertex 4-coloring. ∎

The equivalence of this claim with the topological Four-Color Theorem (Theorem 3.1.3) follows immediately from Theorem 5.1.3, which was proven at the beginning of this chapter.

At this juncture it is prudent to clarify the basic difference between combinatorics and topology—in any case, as we understand it to be in this context. In *combinatorics*, we consider finite sets together with (of necessity, also finite) systems of subsets having no further structure. In *topology*, it is a matter of the geometry of the plane—or of infinite point sets of the plane that are either open or closed or compact and, in addition, continuous mappings between such sets. In topology, the notion of a neighborhood plays a crucial role. We also draw your attention to *algebraic topology*, where one attempts to treat geometric problems with algebraic methods, and to *combinatorial topology*, which we will be using in the future development of our deliberations. We will be combining geometric intuition with combinatorial abstraction, and graphical presentation with formal simplicity.

This means that from now on, we will not be working in a purely combinatorial setting. This would have required a complete translation of the thoughts and ideas of the third and fourth chapters into the language of combinatorics. That would certainly have been possible. One has at one's disposal a multitude of characterizations and properties of planar graphs [WAGNER–BODENDIEK 1990, Chapter 9]. However, the thinking involved in purely formal structures is unequivocally more difficult than when one uses objects in the plane. We will embrace the combinatorial approach mainly because we are looking for minimal criminals of the type that do not have an admissible vertex 4-coloring. In this instance, "minimal" then refers to the number of vertices. In addition, we are restricting ourselves to a particular class of plane graphs. Before we introduce them, we will prove yet another important result about minimal criminals.

Lemma 5.3.2

In a minimal criminal (in the sense of the nonexistence of an admissible vertex 4-coloring), every triangle is the border of a face. ∎

Proof Let $G = (E, \mathcal{L})$ be a minimal criminal and $\mathcal{K} \subset \mathcal{L}$ a triangle in G. First of all, we note that any two given vertices of $\mathcal{K}$ are joined by a unique edge in $\mathcal{K}$ and by no other edge in $\mathcal{L}$. We must show that either the interior domain $I(\mathcal{K})$ or the exterior domain $A(\mathcal{K})$ of $\mathcal{K}$ contains no border points of $\mathcal{L}$.

First we consider $I(\mathcal{K})$. Assume that $\mathbf{x}$ is a border point of $\mathcal{L}$ lying in $I(\mathcal{K})$. Then $\mathbf{x}$ belongs to an edge $B \in \mathcal{L}$ whose interior points must lie entirely in $I(\mathcal{K})$. Since B cannot join two vertices of the triangle $\mathcal{K}$, at least one end point of B must lie in $I(\mathcal{K})$. This implies that if $I(\mathcal{K})$ contains a border point of $\mathcal{L}$, then $I(\mathcal{K})$ must contain a vertex of G. Similarly, it follows that if $A(\mathcal{K})$ contains a border point of $\mathcal{L}$, then $A(\mathcal{K})$ contains a vertex of G.

We assume that not only $I(\mathcal{K})$ but also $A(\mathcal{K})$ contains a vertex of G. Then we construct two graphs G^i and G^a having smaller vertex sets. G^i is formed from G by removing all vertices lying in $A(\mathcal{K})$, as well as all edges incident with at least one of these vertices. G^a is created by removing from G all vertices lying in $I(\mathcal{K})$, as well as all edges incident with at least one of these vertices.

As G, by assumption, is a minimal criminal, both graphs G^i and G^a have a 4-coloring of vertices. The vertices $\mathbf{x}_1$, $\mathbf{x}_2$, $\mathbf{x}_3$ of $\mathcal{K}$ are joined pairwise, not only in G^i but also in G^a, by the edges of $\mathcal{K}$. Thus they must have pairwise distinct colors in each of the colorings. By permuting the colors, we can then find vertex 4-colorings for G^i and G^a such that the vertices $\mathbf{x}_j$ take on the color j for $j = 1, 2, 3$. Both of these colorings, taken together, yield a vertex 4-coloring of G. With this, we have reached the desired contradiction. (Compare a similar argument in the proof of Lemma 3.2.5.) □

Now we are in a position to describe the graphs upon which we, in our search for criminals, would like to center our attention.

Definition 5.3.3
A graph $G = (E, \mathcal{L})$ is said to be *normal* if it is a regular saturated line graph in which every triangle is the boundary of a face.

The edges of a normal graph are, consequently, line segments (Definition 4.2.10), and the bounded faces are bordered by (rectilinear) triangles. The unbounded face is the exterior domain of a triangle (Theorem 4.2.4). The bounded faces whose borders contain the sides of this triangle are pairwise distinct. Otherwise, we would have in total only two faces—contrary to the condition of regularity. A vertex of a graph G is regarded as an *inner vertex* (of G) if it does not belong to the boundary of the unbounded (infinite) face. A complementary notion to this is the term *outer vertex* (of G), which refers to a vertex on the boundary of the unbounded face. To avoid having to distinguish between various possible cases, we note that when considering properties of individual vertices in normal graphs we can always restrict ourselves to inner vertices. If an outer vertex is given, then it certainly does not lie on the boundary of the finite face whose border contains the line segment joining the other two outer vertices. By a double application of the stereographic projection, we can arrange that this face is transformed into the infinite face. In this way, the combinatorial structure of the graph and its normality are retained, but the outer vertex considered originally has become an inner vertex. We note also that normality is actually much too strong an assumption to achieve this benefit. It would have sufficed to assume that for the outer vertex in question there exists a bounded face whose border does not contain it.

Normal graphs are dual to cubic graphs (Theorem 4.5.7). Therefore, if there exist at all minimal criminals with respect to the nonexistence of a vertex 4-coloring, there must be normal graphs among them (Theorems 4.5.5 and 5.1.2, as well as Lemma 5.3.2). Because of this, we can concentrate on examining only normal graphs in any further hunt for minimal criminals. We stress once again (compare page 95) that in spite of the fact that graphs that were not normal have appeared in our discussions, such graphs have always had fewer vertices than the one being tested directly for criminality. So as not to overwork the word "criminal" in the context of vertex colorings, we will adopt the terminology used by Heesch [1969]. From now on, a normal graph that we consider to be a minimal criminal in the sense that it does not permit the existence of a vertex 4-coloring will be called a *minimal triangulation*.

5.4   Rings and Configurations

To prove the Four-Color Theorem, it is necessary to examine in considerable detail certain graphs that crop up primarily as subgraphs of normal graphs. These graphs will be called, in general, "configurations." Their exact definition requires a few preliminary ideas of a technical nature.

Definition 5.4.1
Let $G = (E, \mathcal{L})$ be a graph.
a) A chain $K = (\mathbf{x}_1, \mathbf{x}_2, \ldots, \mathbf{x}_r)$ in G with at least three vertices is said to be *closed* if $\mathbf{x}_1$ and $\mathbf{x}_r$, namely the initial and the final vertices of the chain, are neighbors. In this case, the edge that joins the two end vertices is also considered to be one of the *links* of the chain.
b) A chain $K = (\mathbf{x}_1, \mathbf{x}_2, \ldots, \mathbf{x}_r)$ in G is said to be a *simple closed chain* if it is closed, but $\mathbf{x}_{j_1}$ and $\mathbf{x}_{j_2}$ for $1 < |j_1 - j_2| < r - 1$ are not neighbors.

c) A set R of vertices is called a *ring* if its elements can be arranged so that they form a simple closed chain. In this case, one denotes the number of elements of R also as the *size* of the ring R.

The idea of a ring is regarded as one of Birkhoff's important contributions to the clarification of the Four-Color Problem [BIRKHOFF 1913].

The links of a chain K in a (plane) graph G concatenate to form an arc $B(K)$ *associated with* K. The links of a closed chain K form a circuit $\mathcal{K}$. In this way, we can talk about the closed Jordan curve $J(K)$ *associated with* K that subdivides the plane into the *interior domain* $I(K)$ and the *exterior domain* $A(K)$.

Analogously, a ring R determines a circuit $\mathcal{R}$ and a corresponding closed Jordan curve $J(R)$ whose interior domain (exterior domain) will be denoted by *interior domain* $I(R)$ (*exterior domain* $A(R)$).

On the basis of the definition, a ring always has at least three vertices. Conversely, the vertices of a triangle always form a ring. From this point onwards, we will be using the term triangle in yet a third contextual meaning,[3] namely, to denote a ring with precisely three vertices or a set of three pairwise neighboring vertices of a graph. This is actually the most natural terminology if one takes the literal meanings of the words. Apart from its triangles, normal graphs have many rings. This is illustrated by the following theorem.

Theorem 5.4.2
In a normal graph, the set of neighbors of a vertex always forms a ring whose size is identical to the degree of the vertex. If the vertex is an inner vertex, then it lies in the interior domain of the associated closed Jordan curve and is the only vertex of the graph having this property. ∎

Proof Let $G = (E, \mathcal{L})$ be a normal graph and $\mathbf{y}$ a vertex of G. Set $d = d_G(\mathbf{y})$. Furthermore, denote the d edges of G incident to $\mathbf{y}$ by $B_1, \ldots, B_d$ in such a way that considered as line segments with the common end point $\mathbf{y}$, they are labeled in cyclic order. Let $\mathbf{x}_1, \ldots, \mathbf{x}_d$ be their other end points. Then, for each $j \in \{1, \ldots, d-1\}$, the three vertices $\mathbf{y}$, $\mathbf{x}_j$, and $\mathbf{x}_{j+1}$ belong, respectively, to the boundary of a face of G. This is also true for the three vertices $\mathbf{y}$, $\mathbf{x}_d$, and $\mathbf{x}_1$. As the

---

[3]Compare footnote 1 on page 101.

faces of a normal graph are bounded by triangles, they respectively form the complete set of vertices for each of the face boundaries. In other words, for each $j \in \{1, \ldots, d-1\}$, $\mathbf{x}_j$ and $\mathbf{x}_{j+1}$ are neighboring vertices, as is the case for $\mathbf{x}_d$ and $\mathbf{x}_1$. Thus $K = (\mathbf{x}_1, \ldots, \mathbf{x}_d)$ is a closed chain. Before we prove that the chain is a simple closed chain (part (b) of Definition 5.4.1), we will show the second part of the claim.

For this, we assume that $\mathbf{y}$ is an inner vertex of $\mathbf{G}$. We prove, by contradiction, that $\mathbf{y}$ lies in the interior domain $I(K)$. If $\mathbf{y}$ were to belong to the exterior domain $A(K)$, then the interior points of all edges incident with $\mathbf{y}$ would also belong to $A(K)$. We could then choose a point $\mathbf{z} \in I(K)$ and consider the ray H emanating from $\mathbf{y}$ that contains $\mathbf{z}$. If we start at $\mathbf{y}$ and travel along H, we reach, before we get to $\mathbf{z}$, a point $\mathbf{x}'$ of the closed Jordan curve associated with K. This is the point where we cross over from $A(K)$ to $I(K)$. As $I(K)$ is bounded, by traveling even further, we come to a point $\mathbf{x}'' \in J(K)$, beyond $\mathbf{z}$, at which we again leave $I(K)$. The point $\mathbf{x}''$ cannot be an element of K, that is, $\mathbf{x}''$ cannot be a neighbor of $\mathbf{y}$, because otherwise, the line segment $S = [\mathbf{y}, \mathbf{x}'']$ would be an edge of $\mathbf{G}$ all of whose interior points (as was just noted) would lie in $A(K)$. This contradicts the fact that $\mathbf{z} \in I(K) \cap S$. Therefore, $\mathbf{x}''$ is an interior point of a link B' of K. Without loss of generality, we can assume that the vertices $\mathbf{x}_1$ and $\mathbf{x}_2$ are the end points of B'. The edges B', B_1, and B_2 then form a triangle in $\mathbf{G}$ whose interior contains the point $\mathbf{x}'$ that is a neutral point of the map $\mathcal{L}$. Since $\mathbf{G}$ is, by assumption, a normal graph, the triangle $\{B', B_1, B_2\}$ must be the border of the unbounded face of $\mathbf{G}$ (Definition 5.3.3). It follows that $\mathbf{y}$ cannot then be an inner vertex of $\mathbf{G}$. This provides the required contradiction.

Hence, $\mathbf{y} \in I(K)$. The interior domain $I(K)$ consists of $\mathbf{y}$, the interior points of the edges $B_1, \ldots, B_d$, and the interior domains of the closed Jordan curves $B_j \cup B'_j \cup B_{j+1}$, $j \in \{1, \ldots, d-1\}$, and $B_d \cup B'_d \cup B_1$. Here B'_j for each $j \in \{1, \ldots, d-1\}$ denotes the link of K joining $\mathbf{x}_j$ and $\mathbf{x}_{j+1}$, and B'_d is the link between the end points $\mathbf{x}_d$ and $\mathbf{x}_1$. Thus no other vertices of $\mathbf{G}$ lie in $I(K)$.

The simplicity of K still remains to be proved. For that we can assume that $\mathbf{y}$ is an inner vertex of $\mathbf{G}$ (see page 151) and therefore lies in $I(K)$.[4] Suppose we are given $j_1, j_2 \in \{1, \ldots, d\}$ with $1 < j_1 -$ $j_2 < d - 1$ such that $\mathbf{x}_{j_1}$ and $\mathbf{x}_{j_2}$ are neighbors. Then, from the above description of $I(K)$, the interior points of the edge B joining $\mathbf{x}_{j_1}$ to $\mathbf{x}_{j_2}$ must lie in $A(K)$. The edges B_{j_1}, B, and B_{j_2} together form a closed Jordan curve K' having the property that at least one of the vertices $\mathbf{x}_j$ of K (distinct from $\mathbf{x}_{j_1}$ and $\mathbf{x}_{j_2}$) lies in $I(K')$ and at least one lies in $A(K')$ (Corollary 2.2.10). Hence, the edges B_{j_1}, B, B_{j_2} form a triangle that is not the boundary of a face of $\boldsymbol{G}$. This contradicts the normality of $\boldsymbol{G}$. □

Remarks:

1. We will be extending this result even further. In the next chapter, we will show that in a minimal triangulation, the "secondary neighbors" of a vertex also form a ring (Theorem 6.1.9).

2. Since 0-vertices, 1-vertices, and 2-vertices are excluded in the definition of normal graphs, the previous theorem implies that a ring contains at least three vertices.

3. One can sometimes remove vertices from a closed chain without having lost the property of being closed. In the following diagram not only $(\mathbf{y}_1, \mathbf{y}_2, \mathbf{y}_3, \mathbf{y}_4)$ but also $(\mathbf{y}_1, \mathbf{y}_2, \mathbf{y}_3)$ is a closed chain.

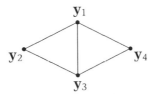

On the other hand, a proper subset of a ring is never a ring in its own right.

Now we are able to introduce the notion mentioned at the beginning of this section.

Definition 5.4.3

A graph $\boldsymbol{C}$ is said to be a *configuration* if

1. it is regular,
2. the outer vertices form a ring whose size is ≥ 4,

---

[4]To those readers who wish to understand this more fully, we recommend that they explicitly consider the case when $\mathbf{y}$ is an outer vertex of $\boldsymbol{G}$.

3. inner vertices exist,
4. the bounded faces have triangular borders,
5. every triangle is the border of a face.

The first of these conditions follows from the remaining four. We have included it as a condition in order to spare ourselves a somewhat long and drawn-out proof. We leave such a proof to the reader.

A nontrivial example of a configuration is the *Birkhoff diamond.*

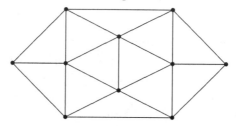

It enjoys as much renown in graph theory as the Kohinoor diamond does in fictional criminal mysteries.

Special, but very simple, configurations are the stars that in the literature are sometimes also called "wheels" [WHITNEY and TUTTE 1972].

Definition 5.4.4
A configuration is called a *star* if it contains only one inner vertex, called its "hub." In particular, it is said to be a *k-star* if it is a star with precisely k outer vertices and therefore $k+1$ vertices ($k \geq 4$) in total.

By the *ring size* of a configuration we mean the size of the ring of its outer vertices. In this meaning of the word, the Birkhoff diamond then has ring size 6, and a k-star has ring size k. The interior domain of the ring of outer vertices of a configuration C we simply call the *interior domain of* C. This topological concept is distinguishable from the combinatorial notion of the *core* of a configuration, which consists of the subgraph spanned by the inner vertices of the configuration. By the subgraph *spanned* by a set of vertices of a graph G we mean the graph consisting of the given vertices and all edges of G that join them. We distinguish between three types of edges:

- *inner edges*, which join two inner vertices,
- *outer edges*, which join two outer vertices,

• *legs*, which join one inner vertex to one outer vertex.

In particular, we speak of a *leg B of the inner vertex* **y** if **y** is the end point of the leg *B*. The subgraph of a configuration *C* that is spanned by the outer vertices of *C* is called the *bounding circuit of C*. Its edges consist of the outer edges of *C*.

Remark: The subdivision of the vertices of a configuration into inner and outer vertices and the classification of the different types of edges are purely combinatorial concepts, meaning that they can be read off solely from the underlying combinatorial graph. Outer edges are precisely those edges that can in only one way be extended to form a triangle. Outer vertices are precisely the end points of outer edges. All other vertices are inner vertices from which the inner edges and the legs are defined. We note that inner edges can be extended to form a triangle in precisely two ways. The reader may wish to muse about how in this way a configuration can be characterized in a purely combinatorial fashion.

Lemma 5.4.5

The boundary of every bounded face of a configuration always contains at least one inner vertex. ■

Proof The bounded faces of a configuration have triangular borders. Since the outer vertices form a ring, three of them can belong to a triangle only if there exists a total of three outer vertices. This, however, is not possible in a configuration. □

The core of a star consists of a single vertex and is, therefore, a "degenerate" graph. In the case of more general configurations as well, the core need not be a regular graph. The diagram on page 155 depicts the core of the Birkhoff diamond. It is not regular, since the unbounded face has two borderlines in common with both of the bounded faces. This, of course, contravenes the definition of regularity (Definition 3.2.8). The following diagram illustrates a configuration whose core consists only of one edge together with its two vertices. Therefore, it contains a final edge, which, by the same token, is not permissible in a regular graph.

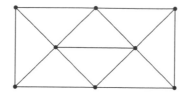

However, one property of regular graphs is always satisfied by the core of a configuration.

Theorem 5.4.6

The core of a configuration is a connected graph.   ■

Proof   Let C be a configuration. We must show that two inner vertices can be joined by a sequence of edges that are not legs or outer edges. Let y_1 and y_2 be two distinct inner vertices of C. We pick, for the time being, an arbitrary arc B joining y_1 to y_2 that lies, however, completely in the interior domain of C. This we can alter so that it lies entirely in the neutrality set of C. To that end, let L be a bounded face of C containing a point of B. Traveling along B from y_1 to y_2, we reach a first point z_1 and a last point z_2 on the boundary of L. We now replace the segment of B between z_1 and z_2 by a polygonal arc in the boundary of L that joins both points but that lies entirely in the interior domain of C. Such an arc exists because the boundary of L contains at least one inner vertex of C (Lemma 5.4.5). By doing this, L in any case no longer contains any point of B. If there exists yet another face of C that contains a point of B, we again similarly alter B. We continue this process until we have achieved our objective. As there are only finitely many faces, this process must terminate after finitely many such alterations.

An arc that lies entirely in the neutrality set of a graph and whose end points are vertices of the graph must contain, along with an interior point of an edge, the entire edge including its end points. Thus neither legs nor outer edges can belong to the arc found above. It therefore consists of a legless sequence of edges joining y_1 to y_2.   □

In the proof of the Four-Color Theorem one considers configurations in a minimal triangulation. We will now make precise exactly what that means. First of all, we must define what it means for two configurations to be "essentially the same." In other words, we need a notion for the equivalence of two configurations.

Definition 5.4.7

Two configurations $C' = (E', \mathcal{L}')$ and $C'' = (E,'' \mathcal{L}'')$ are defined to be *equivalent* if there exists a bijective mapping $\varphi : E' \to E''$ such that it and its inverse preserve the relation of two vertices being neighbors.

The given condition is purely combinatorial. It implies that the underlying graphs are isomorphic to one another. The definition has been kept as technically simple as possible in order that it can be used with ease. For instance, one sees immediately that an equivalence relation is induced on the set of configurations and that all k-stars (for fixed k) are equivalent to one another.

Lemma 5.4.8

Let $C' = (E', \mathcal{L}')$ and $C'' = (E,'' \mathcal{L}'')$ be equivalent configurations. Denote by $\mathcal{G}'$ and $\mathcal{G}''$ the sets of bounded faces of C', respectively C''. Furthermore, suppose that $\varphi : E' \to E''$ is a bijective mapping that represents the equivalence, that is, that preserves the neighbor relation in both directions. Then there exist bijective mappings $\varphi_1 : \mathcal{L}' \to \mathcal{L}''$ and $\varphi_2 : \mathcal{G}' \to \mathcal{G}''$ such that the triple $(\varphi, \varphi_1, \varphi_2)$ of mappings and the triple of their inverses $(\varphi^{-1}, \varphi_1^{-1}, \varphi_2^{-1})$ are compatible with the following incidence relationships: end point of an edge and borderline of a face. These mappings and their inverses also preserve the special classes of vertices and edges. ∎

Proof   First we define $\varphi_1 : \mathcal{L}' \to \mathcal{L}''$. Let $B' \in \mathcal{L}'$ be given. Denote by $\mathbf{z}_1$ and $\mathbf{z}_2$ the end points of B'. Since they are neighboring vertices in C', the vertices $\varphi(\mathbf{z}_1)$ and $\varphi(\mathbf{z}_2)$ are neighboring in C''. Therefore, there exists exactly one edge $B'' \in \mathcal{L}''$ that joins $\varphi(\mathbf{z}_1)$ to $\varphi(\mathbf{z}_2)$. The assignment $B' \mapsto B''$ defines the mapping φ_1 in the desired fashion.

From the construction, it immediately follows that φ_1 maps the outer edges of C' to the outer edges of C''. Thus φ preserves the properties of "outer vertex" and "inner vertex" in both directions. In addition, φ_1 preserves "legs" and "inner edges."

To construct φ_2, we consider $L' \in \mathcal{G}'$. The three vertices on the boundary of L' are pairwise neighboring vertices. At least one of them is an inner vertex (Lemma 5.4.5). Therefore, the images under φ of these vertices are also pairwise neighboring vertices, and at least one of them is also an inner vertex. Since a configuration is regular and each of its triangles borders a face (Definition 5.4.3), these images belong to the boundary of exactly one face L'' of C''. As the boundary of L'' contains an inner vertex, L'' must be a bounded face—that is, $L'' \in \mathcal{G}''$. In this way, the assignment $L' \mapsto L''$ defines the mapping φ_2 as desired.   □

Remark: In particular, two configurations are equivalent if they can be mapped one to the other by a homeomorphism of the plane. From the Wagner and Fáry theorem (Theorem 4.2.11) it follows that every configuration is equivalent to a line graph. We leave to the reader the not-too-difficult proof of the converse: *Equivalent configurations can be mapped one to the other by a homeomorphism of the plane.*

At this point, we are finally ready to define the following fundamental notion.

Definition 5.4.9
The graph G is said to *embed* the configuration C if there exists a closed chain K in G such that the subgraph C_K of G spanned by the vertices of K and the vertices lying in the interior domain of K forms a configuration that is equivalent to C. The configuration C is said to be *properly embedded in G* if K is a simple closed chain.

Remarks: 1. The notion (described above) of an "embedding" of a configuration into a graph will be, for the complete proof of the Four-Color Theorem, generalized even further to the concept of an "immersion." In our presentation, we will not go into these highly technical details.

2. By the *bounding circuit* of a configuration C embedded in a graph G we mean the image of the original bounding circuit. In embeddings that are not proper, there may exist edges that join outer vertices of C in G but that do not belong to C itself. Such edges will not be considered as being part of the bounding circuit of C.

The simplest examples are the k-stars, which are properly embedded in normal graphs (Theorem 5.4.2). Since a normal graph has exactly three outer vertices, it has exactly $v - 3$ stars, where v (> 3) is the number of its vertices. In addition to that, the following holds:

Theorem 5.4.10
A minimal triangulation cannot contain a 4-star but contains at least twelve 5-stars.   ∎

Proof If a normal graph contains a 4-star as a configuration, then the dual map has a country with four neighbors and therefore cannot be a minimal criminal with respect to the nonexistence of a 4-coloring of countries (Theorem 4.5.4). Hence, the original graph cannot be a minimal triangulation either.

The second part of the claim follows directly from inequality (4.6.6). □

We further note that:

Theorem 5.4.11
In a minimal triangulation, the circuit determined by a ring having at least 4 vertices is the bounding circuit of a properly embedded configuration. ■

If one must consider many different configurations and, in doing so, must sketch them, then it is preferable to choose simple presentations. Such simplified presentations were given by Heesch. He described them as *pared images*. His basic premise was that a configuration is essentially determined by its core and the degrees of certain of its vertices. However, first of all, we will formulate a definition for the somewhat difficult exception to this.

Definition 5.4.12
An inner vertex **x** of a configuration C is said to be an *articulation* if the graph that consists of the core of C with **x** and the edges incident with **x** removed from it is no longer connected.

The term "articulation" connotes the idea of a "critical link." This consequently explains its use in this context.[5] The term "cut point" was used previously to denote an articulation. It is somewhat clearer in its meaning.

With the help of the notion of an articulation, the following results can be used to establish the geometric properties of a configuration.

---

[5]The word "articulation" also has the meaning of "the joint or juncture between bones" or "a joint between two separable plant parts."

Lemma 5.4.13

Suppose we are given an articulation of a configuration C. Then the end points of its legs that belong to the ring of outer vertices cannot be arranged to form a chain. This means that the subgraph of C spanned by them is not connected. In particular, an articulation has at least two legs. ∎

Proof Let $\mathbf{y}$ be an articulation of C. Set $d_C(\mathbf{y}) = d$. Denote by $B_1, \ldots, B_d$, in cyclic order, the edges of C that are incident with $\mathbf{y}$. Let $\mathbf{z}_1, \ldots, \mathbf{z}_d$ be the other end points of these edges, respectively. Now we pick two inner vertices $\mathbf{y}_1$ and $\mathbf{y}_2$ of C that are separated by $\mathbf{y}$. In other words, after removing $\mathbf{y}$ (and the outer vertices), they belong to different components of the remaining graph. Furthermore, we choose a legless sequence of abutting edges joining $\mathbf{y}_1$ to $\mathbf{y}_2$. Such a sequence exists, since the core of a configuration is connected (Theorem 5.4.6). Because of the separation property of $\mathbf{y}$, this sequence of concatenated edges must pass through $\mathbf{y}$. Therefore, it must contain two edges having $\mathbf{y}$ as a common end point. Let us call these edges B_1 and B_j, where $j \in \{2, \ldots, d\}$. Then $\mathbf{z}_1$ and $\mathbf{z}_j$ are inner vertices of C, but they are not neighbors of one another, since otherwise, the edge joining them could replace the edges B_1 and B_j in the given edge sequence from $\mathbf{y}_1$ to $\mathbf{y}_2$. This would mean that there exists a legless sequence of concatenated edges from $\mathbf{y}_1$ to $\mathbf{y}_2$ that does not meet $\mathbf{y}$. Therefore, $j \in \{3, \ldots, d-1\}$. A corresponding alteration of the given sequence of edges would also be possible if all vertices $\mathbf{z}_k$, $k \in \{2, \ldots, j-1\}$, or all vertices $\mathbf{z}_l$, $l \in \{j+1, \ldots, d\}$, were inner vertices of C. Hence, we can find k and l such that $\mathbf{z}_k$ and $\mathbf{z}_l$ are outer vertices of C.

Now let r be the ring size of C. Label the outer vertices $\mathbf{x}_1, \ldots, \mathbf{x}_r$ in such a way that $(\mathbf{x}_1, \ldots, \mathbf{x}_r)$ is a (simple closed) chain and that $\mathbf{x}_1 = \mathbf{z}_k$. Then, for one $m \in \{2, \ldots, r\}$, $\mathbf{x}_m = \mathbf{z}_l$. We must show that there exists a $p \in \{2, \ldots, m-1\}$ and a $q \in \{m+1, \ldots, r\}$ such that $\mathbf{x}_p$ and $\mathbf{x}_q$ cannot be neighbors of $\mathbf{y}$.

For that, we assume that no p with the desired property exists. This means that the chain $(\mathbf{x}_1, \ldots, \mathbf{x}_m)$ consists only of neighbors of $\mathbf{y}$. Since the neighbors of $\mathbf{y}$ form a ring (Theorem 5.4.2), the vertices $\mathbf{x}_1, \ldots, \mathbf{x}_m$ must coincide either with the vertices $\mathbf{z}_k, \mathbf{z}_{k+1}, \ldots, \mathbf{z}_l$ or with the vertices $\mathbf{z}_l, \ldots, \mathbf{z}_d, \mathbf{z}_1, \ldots, \mathbf{z}_k$. Then, in this case, either $\mathbf{z}_1$ or

z_j would be an outer vertex of C. This contradicts the fact that z_1 and z_j, by the construction, are both inner vertices of C. The existence of a q with the desired property follows analogously. □

Lemma 5.4.14

If an inner vertex is not an articulation, the end points of its legs that belong to the ring of outer vertices can be arranged into a chain (possibly empty). ∎

Proof Let C be a configuration. Let $\mathbf{y}$ be an inner vertex of C that is not an articulation. Furthermore, let r again denote the ring size of C. Label the outer vertices by $\mathbf{x}_1, \ldots, \mathbf{x}_r$ in such a way that $(\mathbf{x}_1, \ldots, \mathbf{x}_r)$ forms a (simple closed) chain. We assume that the vertices $\mathbf{x}_1$ and $\mathbf{x}_m$ for some $m \in \{2, \ldots, r\}$ are neighbors of $\mathbf{y}$. The corresponding edges linking these vertices to $\mathbf{y}$ will be denoted by B_1 and B_m, respectively. One must show that either all $\mathbf{x}_p$ for $p \in \{1, 2, \ldots, m\}$ or all $\mathbf{x}_q$ with $q \in \{m, m+1, \ldots, r, 1\}$ are neighbors of $\mathbf{y}$. For that, we consider the closed Jordan curves K_+ and K_- associated with the closed chains $(\mathbf{y}, \mathbf{x}_1, \ldots, \mathbf{x}_p, \ldots, \mathbf{x}_m)$ and $(\mathbf{y}, \mathbf{x}_m, \ldots, \mathbf{x}_q, \ldots, \mathbf{x}_r, \mathbf{x}_1)$. If both interior domains were to contain vertices of C, then $\mathbf{y}$ would be an articulation. Therefore, we can assume that no vertex of C can lie in $I(K_+)$. However, the bounded faces to whose borders the links of the chain $(\mathbf{x}_1, \mathbf{x}_2, \ldots, \mathbf{x}_n)$ belong must lie in $I(K_+)$. Since each of these faces has an inner vertex belonging to its border (Lemma 5.4.5), these inner vertices must, therefore, all lie in $I(K_+)$. However, by assumption and by the construction, there is only one inner vertex with this property, namely, the vertex $\mathbf{y}$ under consideration! Since the borders of these faces are triangles, each vertex in the border that is distinct from $\mathbf{y}$ must be a neighbor of $\mathbf{y}$. □

Lemma 5.4.15

Every configuration has inner vertices that have at least two legs but are not articulations. ∎

Proof Let the outer vertices of the configuration C be arranged to form the simple closed chain $K = (\mathbf{x}_1, \ldots, \mathbf{x}_r)$. The pair $(\mathbf{x}_r, \mathbf{x}_1)$ can be complemented by an inner vertex $\mathbf{y}_1$ to form a triangle. If $\mathbf{y}_1$ is not an articulation, then we are finished, since by construction, $\mathbf{y}_1$ has at least two legs. In the case that $\mathbf{y}_1$ is an articulation, we can find indices s_1', s_1'' where $1 \leq s_1' < s_1'' \leq r$ and $s_1'' - s_1' > 1$ such that the vertices $\mathbf{x}_{s'_1}$ and $\mathbf{x}_{s''_1}$ are neighbors of $\mathbf{y}_1$ but none of the other outer vertices between them has $\mathbf{y}_1$ as a neighbor. We now form the closed chain $K_1 = (\mathbf{y}, \mathbf{x}_{s'_1}, \ldots, \mathbf{x}_{s''_1})$. The inner vertex $\mathbf{y}_2$ of C that complements the pair $(\mathbf{x}_{s'_1}, \mathbf{x}_{s'_1+1})$ to form a triangle lies in the interior domain of K_1, the end points of its legs being outer vertices $\mathbf{x}_t$ for some $s'_1 \leq t \leq s''_1$. If $\mathbf{y}_2$ is not an articulation, then we are now finished. Otherwise, we can find indices s'_2, s''_2 with the appropriate properties satisfying the condition that $s''_2 - s'_2 < s''_1 - s'_1$. By continuing this process, the difference $s''_k - s'_k$ can be made ever smaller. However, it cannot be negative. Therefore, the process stops with the discovery of an inner vertex $\mathbf{y}_{k+1}$ having the desired properties. □

The unified representation of configurations is based upon the following theorem.

Theorem 5.4.16

A configuration in which all of its articulations have exactly two legs is determined, up to equivalence, by its core and the degrees of its inner vertices (taken with respect to the entire configuration). ∎

Proof Let C' and C'' be configurations with a common core such that all inner vertices have the same degree with respect to both C' and C'' and such that all articulations have only two legs. We must specify a suitable bijection between the outer vertices of both configurations. To do this, we begin with an inner vertex $\mathbf{y}_1$ that has at least two legs but that is not an articulation. The existence of such a vertex is assured by the previous lemma. On the basis of the above assumptions, $\mathbf{y}_1$ has the same degree d and the same number $s \geq 2$ of legs with respect to C' and to C''. If $s = d$, then C' and C'' are stars of the same ring size and hence are equivalent. We therefore assume that $s < d$. In this case, we can arrange the neighbors of $\mathbf{y}_1$ in C' and C'' into a chain $(\mathbf{z}'_1, \ldots, \mathbf{z}'_d)$, respectively $(\mathbf{z}''_1, \ldots, \mathbf{z}''_d)$, satisfying the following properties: The chains will have the same orientation; the first s vertices of each will be outer vertices; and $\mathbf{z}'_t = \mathbf{z}''_t$ for $t > s$.

In order to proceed with the construction, the following observation is important: *In a configuration, a triangle that contains an outer vertex has exactly two legs in its boundary. The third side can be an outer edge or an inner edge.*

Now we proceed inductively. For $i = 1, 2$, we inductively construct (finite) sequences $(B_k^{(i)})$ of legs and finite sequences $(L_k^{(i)})$ of triangles in $C^{(i)}$ having the legs $B_k^{(i)}$, $B_{k+1}^{(i)}$ as sides.[6] For $k = 1, 2$, let $B_k^{(i)}$ be the leg that joins $\mathbf{y}_1$ with $\mathbf{z}^{(i)}{}_k$. This stipulation is possible because $s \geq 2$. The triangle $L_1^{(i)}$ will be constructed from the vertices $\mathbf{y}_1$, $\mathbf{z}_1^{(i)}$, and $\mathbf{z}_2^{(i)}$. If $L_{k-1}^{(i)}$ and $B_k^{(i)}$ have already been determined, then we take $L_k^{(i)}$ to be the second triangle (distinct from $L_{k-1}^{(i)}$) with the side $B_k^{(i)}$, and we take $B_{k+1}^{(i)}$ to be the second leg (distinct from $B_k^{(i)}$) from the sides of the triangle $L_k^{(i)}$. Since a configuration contains only finitely many triangles, the entries of the sequence $(L_k^{(i)})$ must repeat themselves after some point. The construction shows that there must be indices $p^{(i)}$ such that $L_{p^{(i)}+1}^{(i)} = L_1^{(i)}$. Hence, it also follows that $B_{p^{(i)}+1}^{(i)} = B_1^{(i)}$.

Each of the legs $B_k^{(i)}$ so determined has an inner vertex $\mathbf{y}_k^{(i)}$ and an outer vertex $\mathbf{x}_k^{(i)}$ as its end points. Note that for the first induction step, the inner vertices satisfy the following for $i = 1, 2$:

$$\mathbf{y}_1^{(i)} = \cdots = \mathbf{y}_s^{(i)} = \mathbf{y}_1,$$

whereas the corresponding outer vertices $\mathbf{x}_k^{(i)} = \mathbf{z}_k^{(i)}$ form chains $(\mathbf{x}_1^{(i)}, \ldots, \mathbf{x}_s^{(i)})$ of equal length. The sequences $(\mathbf{x}_1^{(i)}, \ldots, \mathbf{x}_{p^{(i)}}^{(i)})$ are, up to repetitions, (simple) closed chains of outer vertices of $C^{(i)}$. As for the triangles $L^{(i)}$, there are two types: type A and type B. We classify such a triangle as type A if it has a side that is an inner edge; otherwise, the triangle is of type B. The triangles of type B, therefore, have a side that is an outer edge. Now we note that for fixed k, both triangles L_k' and L_k'' always have the same type. Therefore, in the sequences $(L^{(i)})$, the change of type always occurs at the same place $k_j, j \in \{1, \ldots, n\}$, $k_1 \leq \cdots \leq k_n$. Hence, it follows that $p' = p''$. From the construction, the triangles L_1' and L_1'' have an outer edge as a side, and therefore they are of type B. The same holds for the triangles

---

[6]As in differential calculus, the notation $^{(i)}$ for $i = 1$ is generally denoted by a single stroke $'$ and for $i = 2$ a double stroke $''$.

$L_k^{(i)}$, where $k \leq k_1 = s - 1$. Both of the triangles $L_s^{(i)}$, for $i = 1, 2$ are of type A. In addition to the inner vertex $\mathbf{y}_1$, they have another inner vertex $\mathbf{y}_2 = \mathbf{y}_{s+1}^{(i)}$ in common. If $\mathbf{y}_2$ is an inner vertex with only one leg or if it is an articulation, then the triangles $L_{s+1}^{(i)}$ are also of type A and have another inner vertex $\mathbf{y}_3 = \mathbf{y}_{s+2}^{(i)}$ in common. This process continues in a similar fashion until we reach the triangles $L_{k_2}^{(i)}$ for which the vertices $\mathbf{y}'_{\mathbf{k_2}+1} = \mathbf{y}''_{\mathbf{k_2}+1}$ are not articulations but have $s_2 > 1$ legs. Up to this point, the following holds:

$$\mathbf{x}_s^{(i)} = \mathbf{x}_{k_1+1}^{(i)} = \cdots = \mathbf{x}_{k_2+1}^{(i)}.$$

The outer vertices of the legs with inner vertices $\mathbf{y}_{k_2+1}^{(i)}$, $i = 1, 2$, now form chains of equal length:

$$(\mathbf{x}_{k_2+1}^{(i)}, \ldots, \mathbf{x}_{k_2+s_2}^{(i)}).$$

The triangles $L_{k_2+1}^{(i)}, \ldots, L_{k_3}^{(i)}$ for $k_3 = k_2 + s_2 - 1$ are all of type B, after which a change of type again occurs. The obvious continuation of this process yields the desired result. We note that the last triangles $L_{p^{(i)}}^{(i)}$ are both of type A.

    The search described above for the triangles $L_k^{(i)}$ demonstrates the actual goal of this proof. The assignment $\mathbf{x}'_k \mapsto \mathbf{x}''_k$ yields a well-defined bijection from the set of outer vertices of $\mathbf{C}'$ to the set of outer vertices of $\mathbf{C}''$.   □

Remark: The theorem also holds for a weakening of the assumption that the occurring articulations have precisely two legs. One observes that the number of legs of an articulation is always greater than or equal to the number of components into which the core of the configuration subdivides after the articulation and the edges incident with it are removed. Exactly then, when both of these numbers are equal to one another for all of its articulations, is the configuration uniquely determined by its core and the degrees of its inner vertices. Only if an articulation has more legs than corresponding components does one not know how the legs are distributed "among" the components. We will later (page 219) clarify why configurations with articulations having more than two legs can remain outside the realm of consideration for the proof of the Four-Color Theorem.

The *pared image* of a configuration, according to Heesch, is created by drawing the core of the configuration and by specifying the degrees of the vertices, as is illustrated by the symbols in the following table.

Degree	5	6	≥ 6	7	≥ 7	8	9
Symbol	![5]	![6]⁷	![≥6]	![7]	![≥7]	![8]	![9]

On page 155, we have the pared image of the Birkhoff diamond. We now present another two examples of pared images: first of all, the pared image of the configuration on page 158 whose core consists of one edge and two 5-vertices

and, secondly, the pared image

of the configuration shown below

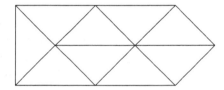

whose core contains one edge, a 5-vertex, and a 6-vertex. Both of these pared images, called the 5-5-chain and the 5-6-chain respectively, were first studied by Paul Wernicke [WERNICKE 1904]. They have a certain historical significance that we will discuss later in the book (see page 224).

---

[7] 6–vertices have no special marking.

5.5   And Now, to the Proof!

In order to show in utmost clarity the significance of configurations for the Four-Color Theorem, we need one more somewhat curious concept.

Definition 5.5.1
A configuration C is said to be *reducible* if a normal graph containing C as a configuration cannot be a minimal triangulation. Otherwise, it is said to be *irreducible*.

These notions are curious precisely because the Four-Color Theorem actually implies that every configuration is reducible. At this point, we must refer back to Heesch's ideas. This is not an "objective" property of a configuration. It is a property that "fluctuates" with respect to its relative position in the progression towards the final proof of the Four-Color Theorem [HEESCH 1969, page 14]. That we can describe different types of reducibility that also remain nontrivial when applied to the Four-Color Theorem is a problem with which we will deal in the next chapter. For the moment, we are able to state that the 4-star is reducible (Theorem 5.4.10). Most unfortunate is the fact that a corresponding argument for normal graphs containing a 5-star as a configuration has, as yet, not been found. If such an argument were to exist, then the Four-Color Theorem would have already been proved, since each normal graph must contain either a 4-star or a 5-star (Corollary 4.6.3). Pity! However, this situation points the way to the final proof of the Four-Color Theorem.

Definition 5.5.2
A set $\mathcal{U}$ of configurations is said to be *unavoidable* if each normal graph contains an element of $\mathcal{U}$.

The discussion preceding this definition implies, therefore, that a set consisting of a 4-star and a 5-star is unavoidable. The ideas of Paul Wernicke mentioned in the previous section suggest that the set consisting of the 4-star together with the two configurations whose pared images are shown on page 167 is also unavoidable. With the methods developed by Heesch, the proof of unavoidability is much easier (see page 223) than if one tries to use Wernicke's arguments [WERNICKE 1904]. Unfortunately, however, both of these configurations are far from being reducible.

With the notions of "reducible configuration" and "unavoidable set of configurations" at one's disposal, a way can be found that allows one to proceed towards the final proof of the Four-Color Theorem. This way had actually been mapped out since Birkhoff's work in 1913: *"One constructs an unavoidable set of reducible configurations!"* This means that every normal graph must contain an element of this set and therefore cannot be a minimal triangulation. If, however, no minimal triangulation exists, then the Four-Color Theorem must be true.

6 Reducibility

CHAPTER

6.1 Kempe Chain Games

One method essential to proving the reducibility of certain configurations goes back to Kempe and therefore is called the *Kempe chain game*. In this section we wish to discuss a couple of examples that illustrate this procedure. To do so, we will require a few preparatory remarks.

Definition 6.1.1
A *colored graph* is a pair (G, χ) consisting of a (plane) graph G and an admissible vertex 4-coloring χ of G.

If we are given a colored graph (G, χ), then for each pair b, g of colors, $b, g \in \{1, 2, 3, 4\}$, $b \neq g$, we denote by G_{bg} the subgraph of G that is spanned by all vertices having colors b and g.

Definition 6.1.2
Let (G, χ) be a colored graph.
(a) A *Kempe chain* is a chain whose vertices are colored with only two colors. In particular, we speak of a (b, g)-*chain* if it is a Kempe chain whose vertices are (alternately) colored with the colors b, $g \in \{1, 2, 3, 4\}$.

(b) A *Kempe net* is a component of a subgraph of the form G_{bg}. In particular, we speak of a (b, g)-*net* if it is a Kempe net whose vertices are colored with the colors $b, g \in \{1, 2, 3, 4\}$.

Each pair of vertices of a Kempe net can be joined by a Kempe chain. In fact, Kempe nets can be characterized using Kempe chains.

Lemma 6.1.3
If (G, χ) is a colored graph, then the subgraph $C = (E_C, \mathcal{L}_C)$ of G is a Kempe net if and only if:
1. *The vertices of C are completely colored using only two colors.*
2. *Each pair of vertices of C can be joined by a Kempe chain all of whose vertices belong to C.*
3. *E_C is maximal with respect to properties 1 and 2.*
4. *C is spanned by E_C.* ■

We are now ready to describe the moves in a Kempe chain game.

Definition 6.1.4
Let (G, χ) be a colored graph. The coloring $\tilde{\chi}$ of $G = (E, \mathcal{L})$ is produced from χ by a *Kempe interchange* if the colors in a Kempe net are interchanged, that is, if a (b, g)-net $C = (E_C, \mathcal{L}_C)$ exists such that

$$\tilde{\chi}(\mathbf{z}) = \begin{cases} g & \text{for } \mathbf{z} \in E_C \text{ and } \chi(\mathbf{z}) = b, \\ b & \text{for } \mathbf{z} \in E_C \text{ and } \chi(\mathbf{z}) = g, \\ \chi(\mathbf{z}) & \text{for } \mathbf{z} \in E \setminus E_C. \end{cases}$$

A vertex 4-coloring that is produced from an admissible coloring by a Kempe interchange is itself admissible. The fundamental significance of the Kempe interchange is based upon the following facts, which are of a very technical nature.

Lemma 6.1.5
Let $K = (\mathbf{x}_1, \ldots, \mathbf{x}_r)$, $r \geq 4$, be a closed chain in a colored graph (G, χ) whose associated closed Jordan curve $J(K)$ borders a face of G. Suppose, in addition, that the vertices $\mathbf{x}_1$ and $\mathbf{x}_j$ for some j such that $2 < j < r$ belong to the same $(1, 2)$-net. Then there exist two vertices $\mathbf{x}_k$ and $\mathbf{x}_l$, $1 < k < j, j < l \leq r$, bearing the colors 3 and 4 that do not belong to the same $(3, 4)$-net. ■

In this situation, one can alter the color of $\mathbf{x}_k$ from 3 to 4 (or in reverse) by a Kempe interchange without having to recolor $\mathbf{x}_l$.

Proof Denote by L the face bounded by $J(K)$. We can assume, without loss of generality, that $L = I(K)$. Using the Schoenflies theorem, we can further assume that $J(K)$ is the unit circle and that the points $\mathbf{x}_1$, $\mathbf{x}_k$, $\mathbf{x}_j$, and $\mathbf{x}_l$ are vertices of a square Q inscribed within the unit circle. Thus we can also interpret Q as a closed Jordan curve.

The assumption yields a $(1, 2)$-chain K' from $\mathbf{x}_1$ to $\mathbf{x}_j$ such that the associated arc $B(K')$, up to its end points, lies entirely in the exterior domain $A(Q)$. We extend $B(K')$ by the inclusion of a suitable diagonal of Q to form a closed Jordan curve G'. Since a subarc of G' lies in $I(Q)$ and the complementary subarc lies in $A(Q)$ (respectively, up to its end points), one of the vertices $\mathbf{x}_k$, $\mathbf{x}_l$ lies in $I(G')$ and the other in $A(G')$ (Corollary 2.2.10).

We now consider a chain K'' from $\mathbf{x}_k$ to $\mathbf{x}_l$. The associated arc $B(K'')$ must intersect G' and lies, except for its end points, entirely in $A(Q)$. Therefore, it must intersect the arc $B(K')$. Now, the arcs $B(K')$ and $B(K'')$ are unions of edges of $\mathbf{G}$. Hence, if their intersection is nonempty, it consists of isolated vertices and entire edges of $\mathbf{G}$ (inclusive of the end points). The chains K' and K'' therefore have at least one vertex in common, implying that K'' contains one vertex colored with either color 1 or color 2. This means that K'' is not a $(3, 4)$-chain. Since this is true for each chain from $\mathbf{x}_k$ to $\mathbf{x}_l$, the vertices $\mathbf{x}_k$ and $\mathbf{x}_l$ do not lie in the same $(3, 4)$-net. □

Remark: One could ask why in the above proof the introduction of the square Q was necessary at all. Could one instead not have been able to work directly with the closed Jordan curve $J(K)$? That would not have been possible, since one of the interior vertices of the chain K' may also have been a vertex of the closed chain K. In that case, the conditions for Corollary 2.2.10 would not have been fulfilled, and the corollary could not have been applied.

As a first application, we show once again that the 4-star is reducible (compare Theorem 5.4.10).

Lemma 6.1.6

There can be no 4-vertex in a minimal triangulation. ∎

Proof Let $\mathbf{y}$ be a 4-vertex in a minimal triangulation $\mathbf{G}$. Denote by $\mathbf{x}_1, \ldots, \mathbf{x}_4$ the neighbors of $\mathbf{y}$ in cyclic order. The graph $\mathbf{G}'$ is then formed from $\mathbf{G}$ by removing the vertex $\mathbf{y}$ and the edges incident with **y**. Since **G** is, by assumption, a minimal triangulation, there exists an admissible 4-coloring of vertices for **G′**. We choose such a coloring, denote it by χ', and form the colored graph (G', χ'). If only three colors are needed for coloring the neighbors of **y** contained in **G′**, then we have a fourth color free for **y**. We can then extend χ' directly to an admissible vertex 4-coloring of **G**.

If the neighbors of **y** are pairwise differently colored, then by a permutation of colors we can achieve that

$$\chi'(\mathbf{x}_i) = i$$

for all $i \in \{1, 2, 3, 4\}$. It is at this point that the actual Kempe chain game begins.

Move 1: If $\mathbf{x}_1$ and $\mathbf{x}_3$ belong to different $(1, 3)$-nets, then we perform a Kempe interchange in the $(1, 3)$-net to which $\mathbf{x}_1$ belongs. After that, $\mathbf{x}_1$ has color 3, implying that the new coloring of **G′** needs only the three colors 2, 3, and 4 for the neighbors of **y**. We are now able to extend this coloring to all of **G**. With this, the game is finished. In the case that $\mathbf{x}_1$ and $\mathbf{x}_3$ belong to the same $(1, 3)$-net, one must move again.

Move 2: If the vertices $\mathbf{x}_1$ and $\mathbf{x}_3$ belong to the same $(1, 3)$-net, $\mathbf{x}_2$ and $\mathbf{x}_4$ must then belong to distinct $(2, 4)$-nets (Lemma 6.1.5). By a Kempe interchange, we can now obtain a coloring χ'' satisfying

$$\chi''(\mathbf{x}_2) = \chi''(\mathbf{x}_4) = 4$$

that requires only the three colors 1, 3, and 4 for the neighbors of **y**. This coloring can be extended to all of **G**, and the game is finally over. □

In this same way, Kempe also tried to prove the nonexistence of a 5-star in a minimal triangulation. Kempe's arguments and the fallacy contained therein are so typical for the subsequent proof of the Four-Color Theorem that we wish to present both of them here.

Claim: *In a minimal triangulation, there exists no 5-vertex.*

Attempt at a Proof Let **y** be a 5-vertex in a minimal triangulation **G**. Denote the neighbors of **y** by $\mathbf{x}_1, \ldots, \mathbf{x}_5$ in cyclic order. Suppose that the graph **G′** is formed from **G** by removing the vertex **y** and the edges incident with **y**. Since **G** is a minimal triangulation, there exists an admissible 4-coloring of vertices of G'. We choose one such coloring, denote it by χ', and consider the colored graph (G', χ'). We may limit ourselves to the examination of those cases in which all four colors are required for the coloring of the neighbors of $\mathbf{y}$. By renumbering and permutation, we are able to achieve the following color distribution:

$$\chi'(\mathbf{x_i}) = \begin{cases} i & \text{if} \quad i \in \{1, 2, 3, 4\}, \\ 2 & \text{if} \quad i = 5. \end{cases}$$

Now the Kempe chain game begins again.

Move 1: If $\mathbf{x}_1$ and $\mathbf{x}_3$ lie in different $(1, 3)$-nets, using a Kempe interchange we recolor so that $\mathbf{x}_1$ takes on color 3 and color 1 is free for $\mathbf{y}$.

Move 2: If $\mathbf{x}_1$ and $\mathbf{x}_4$ lie in different $(1, 4)$-nets, again using a Kempe interchange, we recolor so that $\mathbf{x}_1$ takes on color 4 and color 1 is free for $\mathbf{y}$.

Move 3: If there exists a $(1, 3)$-chain K_3 from $\mathbf{x}_1$ to $\mathbf{x}_3$ and a $(1, 4)$-chain K_4 from $\mathbf{x}_1$ to $\mathbf{x}_4$, then there exists neither a $(2, 4)$-chain from $\mathbf{x}_2$ to $\mathbf{x}_4$ nor a $(2, 3)$-chain from $\mathbf{x}_5$ to $\mathbf{x}_3$. We are able to recolor so that $\mathbf{x}_2$ takes on color 4 and $\mathbf{x}_5$ can be colored with color 3. Now color 2 is free for $\mathbf{y}$, and the claim is proven. $\qquad\square$

So, where is the error? It took more than ten years before it was found by Heawood. The secret lies hidden in move 3. Since two re-colorings occur simultaneously, one of them must follow the other. It can then happen that in carrying out the first recoloring, the conditions under which one can carry out the second are destroyed. This is certainly not the case if the $(1, 3)$-net containing $\mathbf{x}_1$ and the $(1, 4)$-net containing $\mathbf{x}_1$ have no common vertex other than $\mathbf{x}_1$. However, apart from that, in simultaneous recolorings, an inadmissible coloring can arise. One possible scenario is illustrated by the following diagram in which we have given the vertices the colors supplied by χ'. (So as not to make the diagram too large but nevertheless to draw the lines as far apart as possible, we have drawn not only line segments but also curves. It is left to the reader to devise a suitable transformation to a line graph.)

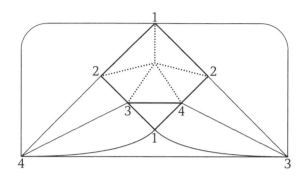

A map dual to the graph pictured above is shown on the front cover of this book. This counterexample is essentially simpler than the original one offered by Heawood [HEAWOOD 1890]. It was Saaty who presented the graph dual to Heawood's map [SAATY 1972, page 9].

Using Kempe chain games, Birkhoff showed that all configurations of ring size 4 as well as all configurations of ring size 5 with more than one inner vertex are reducible. We do not wish to discuss these particular results here. Instead, we refer interested readers to the original paper [BIRKHOFF 1913] and to Aigner's book [1984, Theorems 9.1 and 9.2]. As an immediate corollary, we have the following result.

Theorem 6.1.7
In a minimal triangulation, there exists neither
1. *a ring of size 4, nor*
2. *a ring of size 5 whose interior domain contains more than one vertex of* **G**.   ∎

Thus, a minimal triangulation contains only relatively few types of closed chains of short length. The nonsimple closed chains of length ≤ 5 together with the connecting edges and a typical interior are shown in the following diagram.

The only simple closed chains of length ≤ 5 or rings of size ≤ 5, besides the 3-chains consisting of the vertices of a triangle, are formed by the neighbors of 5-vertices. This situation led to the following definition [ROBERTSON et al. 1997].

Definition 6.1.8

A normal graph is called an *internally 6-connected triangulation* if every ring of size ≤ 5 consists of either the vertices of a triangle or the neighbors of a 5-vertex.

With this notion, the preceding theorem means that a minimal triangulation is internally 6-connected. Therefore, in their new proof of the Four-Color Theorem [ROBERTSON et al. 1997], Robertson, Sanders, Seymour, and Thomas restricted their attention to interally 6-connected triangulations.

Another theorem that Birkhoff proved using the above results is easy to establish. It extends the claim that the neighbors of a vertex form a ring (Theorem 5.4.2), namely, that the "secondary neighbors" of a vertex also form a ring.

Theorem 6.1.9

Let $\mathbf{y}$ be a vertex of a minimal triangulation $\mathbf{G}$. Let R be the set of all vertices of $\mathbf{G}$ that are distinct from $\mathbf{y}$ and its neighboring vertices but that are neighbors of a neighboring vertex of $\mathbf{y}$. Then R is a ring with at least $d_\mathbf{G}(\mathbf{y})$ vertices. ∎

Proof: We set $d = d_\mathbf{G}(\mathbf{y})$ and denote the neighbors of $\mathbf{y}$ by $\mathbf{x}_j, j \in \{1, \ldots, d\}$, so that $(\mathbf{x}_1, \ldots, \mathbf{x}_d)$ forms a (simple) closed chain. Without loss of generality, we can assume that $\mathbf{y}$ is an inner vertex of $\mathbf{G}$ (see page 151). We let $B_1, \ldots, B_d$ denote the links of this chain in such a way that

$$B_j = \begin{cases} [\,\mathbf{x}_d \quad, \mathbf{x}_1\,] & \text{for } j = 1, \\ [\,\mathbf{x}_{j-1}, \mathbf{x}_j\,] & \text{otherwise.} \end{cases}$$

Each vertex $\mathbf{x}_j$ is of degree greater than or equal to 5 and thereby is an end point of at least two edges whose other end points belong to R. We denote these edges by B_{jk_j}, $k_j \in \{1, \ldots, e_j\}$, where $e_j = d_\mathbf{G}(\mathbf{x}_j) - 3$. Indeed, this is done in such a way that the edges $B_j, B_{j1}, \ldots, B_{je_j}, B_{j+1}$ for $j \in \{1, \ldots, d-1\}$ and the edges $B_d, B_{d1}, \ldots, B_{de_d}, B_1$ (the case when $j = d$) are cyclically ordered in the neighborhood of $\mathbf{x}_j$. Now we note that G divides the plane into triangles. On the one hand, this implies that the end points belonging to R of both edges B_{je_j} and B_{j+11} for $j \in \{1, \ldots, d-1\}$ as well as the end points of the edges B_{de_d} and B_{11} coincide. On the other hand, the end points of both edges B_{jk} and $B_{jk+1}, j \in \{1, \ldots, d\}, k \in \{1, \ldots, e_j - 1\}$, are neighbors of each other. For $j \in \{1, \ldots, d\}$ and $k \in \{1, \ldots, e_j - 1\}$, we denote by $\mathbf{z}_{jk}$ the end points of B_{jk} belonging to R. In doing so, we have covered all vertices belonging to R. From this, it follows that the vertex $\mathbf{z}_{11}$ is also an end point of B_{de_d} and, for $j \in \{2, \ldots, d\}$, the vertex $\mathbf{z}_{j1}$ is also an end point of $B_{j-1e_{j-1}}$. Moreover, in the sequence

$$S = (\mathbf{z}_{11}, \ldots, \mathbf{z}_{1\,e_1-1}, \mathbf{z}_{21}, \ldots, \mathbf{z}_{d\,e_d-1}),$$

each pair of adjacent vertices including the pair of the first and the last vertices are neighbors. First of all, we claim that the sequence S consists entirely of elements of the set R. By the construction of S, the elements of R certainly all crop up in the sequence S. To prove the converse, one must show that no element of S is a neighbor of $\mathbf{y}$, that is, that

$$\mathbf{z}_{j_1 k} \neq \mathbf{x}_{j_2}$$

holds for all possible 3-tuples (j_1, j_2, k) of indices. The case when $j_1 = j_2$ follows again directly from the construction. If there were to exist a 3-tuple of indices (j_1, j_2, k) where $j_1 \neq j_2$ and

$$\mathbf{z}_{j_1 k} = \mathbf{x}_{j_2},$$

then $\mathbf{x}_{j_1}$ and $\mathbf{x}_{j_2}$ would be neighbors. Since the vertices $\mathbf{x}_j$ form a ring, the edge $B_{j_1 k}$ must coincide with one of the edges B_j. This, from the construction, is not possible. In particular, it follows that the vertices $\mathbf{z}_{jk}$ all lie in the exterior domain of the closed Jordan curve formed by the edges B_j. Note that from the outset, we had assumed $\mathbf{y}$ to be an inner vertex of G.

To show that the sequence S is a closed chain, it suffices to prove that the elements of this sequence are pairwise distinct. This is certainly the case for pairs of vertices $\mathbf{z}_{j_1 k_1}, \mathbf{z}_{j_2 k_2}$ where $j_1 = j_2$ and $k_1 \neq k_2$. We now assume that

$$\mathbf{z}_{j_1 k_1} = \mathbf{z}_{j_2 k_2}$$

where $j_1 \neq j_2$. There are two cases to consider:

1. The vertices $\mathbf{x}_{j_1}$ and $\mathbf{x}_{j_2}$ are neighbors. Then we can, without loss of generality, assume that $j_2 = j_1 + 1$. In that case, $\{\mathbf{x}_{j_1}, \mathbf{x}_{j_2}, \mathbf{z}_{j_1 k_1} = \mathbf{z}_{j_2 k_2}\}$ forms a ring of size 3. Since a normal graph is presumed, the triangle $\{B_{j_2}, B_{j_2 k_2}, B_{j_1 k_1}\}$ is the boundary of a face. As the edge B_{j_2} belongs to the boundary of this face, it follows that $k_2 = 1$ and $k_1 = e_{j_1}$. A vertex $\mathbf{z}_{jk}$ where $j = j_1$ and $k = e_{j_1}$ does not exist.

2. The vertices $\mathbf{x}_{j_1}$ and $\mathbf{x}_{j_2}$ are not neighbors. Then, without loss of generality, we can assume that $j_1 = 1$ and $2 < j_2 < d$, implying that $\{\mathbf{y}, \mathbf{x}_1, \mathbf{z}_{1 k_1}, \mathbf{x}_{j_2}\}$ is a ring of size 4. This, however, is not possible in a minimal triangulation (Theorem 6.1.7, part 1).

Therefore, S is a closed chain. Moreover, since $e_j \geq 2$ for all j, the following is true:

$$\sum_{j=1}^{d} (e_j - 1) \geq d.$$

What remains to be shown is the simplicity of the chain S. As the vertices $\mathbf{z}_{jk}$ for fixed j belong to the ring of neighbors of $\mathbf{x}_j$, we need only show that any pair of vertices $\mathbf{z}_{j_1 k_1}$ and $\mathbf{z}_{j_2 k_2}$ are not neighboring vertices for the case when $j_1 < j_2$ and

- in the case that $j_2 = j_1 + 1$: $k_1 \neq e_{j_1} - 1$ or $k_2 \neq 1$;
- in the case that $j_1 = 1, j_2 = d$: $k_1 \neq 1$ or $k_2 \neq e_d - 1$.

We assume that two such vertices are given to be neighbors. Again, there are the two distinct cases mentioned above to consider:

1. $j_2 = j_1 + 1$: Since the elements of the sequence S are pairwise distinct, $(\mathbf{x}_{j_1}, \mathbf{z}_{j_1 k_1}, \mathbf{z}_{j_2 k_2}, \mathbf{x}_{j_2})$ is a ring of size 4. Such a ring, however, cannot exist in a minimal triangulation (Theorem 6.1.7, part 1).

2. $j_1 = 1$ and $2 < j_2 < d$: Now $(\mathbf{y}, \mathbf{x}_1, \mathbf{z}_{1 k_1}, \mathbf{z}_{j_2 k_2}, \mathbf{x}_{j_2})$ is a simple closed chain of five vertices. Thus we have a configuration of ring size 5 whose interior domain contains at least two vertices of G, namely, either $\mathbf{x}_2$ and $\mathbf{z}_{21}$ or $\mathbf{x}_d$ and $\mathbf{z}_{d1}$. This situation is again not possible (Theorem 6.1.7, part 2).

At this juncture, we would like to present another of Birkhoff's results that is particularly nice—especially since it illustrates another version of the Kempe chain game. The number of possible moves will be extended to include contractions (Definition 5.2.8).

Theorem 6.1.10

A minimal triangulation cannot contain a Birkhoff diamond. ∎

Proof   Let **G** be a minimal triangulation that we assume contains a Birkhoff diamond. We will use the notation in the following diagram.

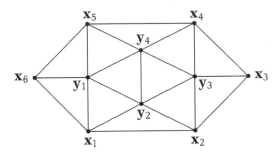

The graph **G'** is formed from the graph **G** by removing the inner vertices of the Birkhoff diamond and the edges incident to them. From this, a face results that is bounded by a hexagon. The Kempe chain game now begins. The first move in this game is not a Kempe interchange. It is a contraction.

Move 1: We form a graph **G''** from the graph **G'** in the following way:
(i) We extend **G'** by adding an edge, namely, the diagonal that joins the vertices x_2 and x_4.

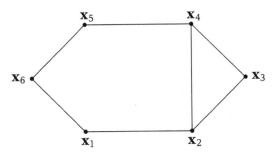

To ensure the possibility of such an extension, we must consider whether the vertices x_2 and x_4 are not already linked by an edge of **G'**. This, however, cannot be the case, since otherwise, the vertices x_2, x_4, and y_3 would have formed a triangle in **G** that would not have been the boundary of a face. This contradicts the definition of a minimal triangulation.
(ii) Then we contract the new edge $[x_2, x_4]$ to the point y_3 lying in it (see page 148/9).

(iii) Finally, we add the line segment from $\mathbf{x}_6$ to $\mathbf{y}_3$ as an edge. To do this, we must be sure that these vertices are not already joined by an edge. However, this can certainly not occur, since otherwise, either the vertices $\mathbf{x}_2$ and $\mathbf{x}_6$ or the vertices $\mathbf{x}_4$ and $\mathbf{x}_6$ must be neighbors in G, and that would then yield, in either case, a ring of size 4 in G. This contradicts the fact that there exist no rings of size 4 in minimal triangulations (Theorem 6.1.7, part 1).

From all of this, the following diagram emerges.

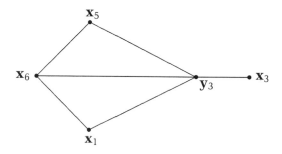

In the framework of a general theory, we will later refer to this diagram as a *reducer* for the Birkhoff diamond (see page 208). As one observes, it is not a configuration.

The graph G'' has five vertices fewer than G. Therefore, it has an admissible 4-coloring χ'' of its vertices. For the colors, we will be using the numbers 0, 1, 2, 3. The vertices of the triangle $[\mathbf{x}_1\mathbf{y}_3\mathbf{x}_6]$ must have three distinct colors. Since if necessary we can alter χ'' by a permutation of colors, we may assume that $\chi''(\mathbf{x}_1) = 0$, $\chi''(\mathbf{y}_3) = 1$, and $\chi''(\mathbf{x}_6) = 2$. For the vertices $\mathbf{z}$ of G', we set

$$\chi'(\mathbf{z}) = \begin{cases} 1 & \mathbf{z} \in \{\mathbf{x}_2, \mathbf{x}_4\}, \\ \chi''(\mathbf{z}) & \text{otherwise.} \end{cases}$$

We thus obtain an admissible 4-coloring of the vertices of G'. In this way, the vertices $\mathbf{x}_2$ and $\mathbf{x}_4$ have the same color. This is permissible, since as has already been observed, the vertices $\mathbf{x}_2$ and $\mathbf{x}_4$ are not neighbors in G'.

In regard to the outer vertices of the Birkhoff diamond, the following six cases are possible.

	x_1	x_2	x_3	x_4	x_5	x_6
1.	0	1	0	1	3	2
2.	0	1	2	1	3	2
3.	0	1	3	1	3	2
4.	0	1	2	1	0	2
5.	0	1	3	1	0	2
6.	0	1	0	1	0	2

Move 2: In the first five cases, we can extend χ' directly to an admissible vertex 4-coloring of G. This is illustrated in the following table (also done without Kempe interchanges).

	y_1	y_2	y_3	y_4
1.	1	3	2	0
2.	1	2	3	0
3.	1	3	2	0
4.	1	2	0	3
5.	1	2	0	3

Move 3: When the boundary of the Birkhoff diamond is colored by χ' in accordance with case 6, a direct extension of χ' to G is not possible.

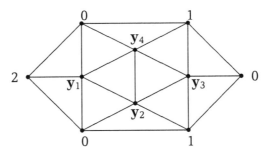

The vertices y_2 and y_4 must be colored with the colors 2 and 3. Thus no color remains at our disposal to color y_3. We now consider different possibilities for the $(0, 3)$-net C to which the vertex x_5 belongs:

i. If the vertex x_3 also belongs to C, then, by a Kempe interchange, we obtain a coloring $\tilde{\chi}'$ where

$$\tilde{\chi}'(x_4) = 2$$

(Lemma 6.1.5). This coloring can be extended in exactly one way to all of G, namely, by the following assignment:

$$
\mathbf{y}_j \mapsto \begin{cases} 3 & \text{for } j = 1, \\ 2 & \text{for } j = 2, \\ 3 & \text{for } j = 3, \\ 1 & \text{for } j = 4. \end{cases}
$$

ii. If the vertex $\mathbf{x}_3$ does not belong to C but the vertex $\mathbf{x}_1$ does, then $\mathbf{x}_1$ and $\mathbf{x}_3$ lie in different $(0, 3)$-nets. By a Kempe interchange, we obtain a coloring $\tilde{\chi}'$ where

$$
\tilde{\chi}'(\mathbf{x}_3) = 3 \, .
$$

This is essentially the coloring of case 5 in move 1. Case 5 in move 2 shows how the coloring can be directly extended to all of G.

iii. If the vertices $\mathbf{x}_1$ and $\mathbf{x}_3$ both do not belong to C, then by a Kempe interchange we obtain a coloring $\tilde{\chi}'$ where

$$
\tilde{\chi}'(\mathbf{x}_5) = 3 \, .
$$

This is case 1 of move 1. Case 1 of move 2 directly extends the coloring to all of G. With this, the game is over. $\qquad\square$

The Birkhoff diamond involves a configuration of ring size 6. Arthur Bernhart studied the reducibility of such configurations thoroughly [BERNHART 1947]. His results, together with those of Birkhoff, form the foundation of the work of Appel and Haken [APPEL and HAKEN 1989, page 171]. Birkhoff himself had already noticed that a configuration of ring size 6 with least four inner vertices and whose inner vertices all have degree 5 contains the Birkhoff diamond. Therefore, such a configuration cannot occur in a minimal triangulation.

6.2   The Birkhoff Number and the Historical Development of Reducibility

In recognition of Birkhoff's contribution to the Four-Color Problem, Saaty introduced the notion of the "Birkhoff number" [SAATY 1972, page 26]. Saaty's definition,[1] however, has been superseded in the meantime by the successful solution of the Four-Color Problem. This is the reason why this notion today is defined differently from its original definition. By the Birkhoff number one means a size that is dependent on the date on the calendar. For each year between 1852 and 1976, it is a measure of the respective progress on the Four-Color Problem. The Birkhoff number is b on day t if on day t it has been shown that a map that cannot be admissibly colored with four colors must contain at least b countries. For the years 1852 to 1879, by the Weiske theorem (Theorem 4.5.1), the Birkhoff number was 6 (Corollary 4.5.2). In 1879, by way of Kempe's deliberations, the Birkhoff number had risen to 13. Birkhoff himself did not improve this estimate. The origin of mathematical results and the subsequent crediting for the results here have—for which there also exist many other instances—not much to do one with the other.

Philip Franklin, in 1922, succeeded in doubling the Birkhoff number—meaning that it rose to 26. In anticipation of the presentation of the historical development, the progression of the Birkhoff number is illustrated below.

---

[1]Saaty had defined the *Birkhoff number* to be the minimum number of countries that a map must have if it cannot be colored admissibly with four colors. By this definition, one must now either establish that the Birkhoff number does not exist at all or assign it the value ∞ (infinity) in accordance with a common convention.

	t	b
B.G. Weiske	before 1852	6
A.B. Kempe	1879	13
P. Franklin	1922	26
C.N. Reynolds	1926	28
P. Franklin	1938	32
C.E. Winn	1940	36
O. Ore and J.G. Stemple	1968	41
W.R. Stromquist	July 2 1973	45
J. Mayer	September 1973	48
W.R. Stromquist	spring 1974	52
J. Mayer	1975	96

As already noted earlier (see Section 5.5), the final proof of the Four-Color Theorem does not heavily depend on the Birkhoff number. From 96 to ∞ is still a good distance to go. The proof depends, however, on the further development of the theory of reducibility begun by Birkhoff, the refinement of discharging procedures (discussed in Chapter 7), the creation of unavoidable sets of configurations, and, above all, an appropriate means of visualization. All of these are areas in whose development Heesch played a major role. We will list here—without proof—the significant results on reducibility in chronological order of achievement.

In a minimal triangulation, the following cannot occur:
- **Birkhoff 1913** a vertex all of whose neighbors have degree 5.
- **Birkhoff 1913** a vertex of even degree all of whose neighbors have degree 6.
- **Franklin 1922** a 5-5-5-chain whose vertices are neighbors of one and the same 6-vertex.
- **Franklin 1922** a 5-vertex whose neighbors can be arranged into a 5-5-6-6-6-chain.
- **Errera 1925** only vertices of degree 5 and 6.
- **Winn 1937** a 5-vertex all of whose neighbors have degree 6.
- **Chojnacki–Hanani 1942** only vertices with degrees not equal to 6 or 7.
- **Heesch 1958** only vertices of degrees 5 and 7, but without a 7-7-7-triangle.

- **Stanik 1973**   only vertices having degrees not equal to 6, but without a 5-5-5-triangle.
- **Osgood 1974**   only vertices with degrees 5, 6 and 8.
- **Allaire 1976**   only vertices with degrees not equal to 6.

6.3   Types of Reducibility — A General Treatment

We are now able to formulate precisely the notions of reducibility that are no longer dependent on the respective progress of the Four-Color Theorem. Heesch, first of all, distinguished between A-, B-, C- and D-reducibility [HEESCH 1969]. Later, he developed other reduction techniques, which however, were not utilized by Appel, Haken, and Koch in the proof of the Four-Color Theorem. It is conceivable that they could have been used to form the foundation for an alternative proof [BIGALKE 1988, pages 229, 239, 255f., 311f.], [HEESCH 1974]. At any rate, the alphabetical sequence used for reducibility types is not as arbitrary as it first appears: A is in honor of A. Errera, B signifies Birkhoff, and C refers to C.E. Winn. This time Birkhoff's name comes in the correct order. His treatment of the diamond, also named after him, is a prototype of B-reduction.

D-reducibility was discovered by Heesch himself and was so called in order to preserve the alphabetical ordering. It is based solely upon Kempe chain games. It also does not use contractions in the way that they were used in the proof of the reducibility of the Birkhoff diamond (see page 180/1). Heesch, too, had noted that D-reducibility could be established, in principle, by finitely many doable calculations using a computer. Karl Dürre was the first to undertake a test run on the computer in regard to these calculations (see page 27). We will present the Dürre–Heesch algorithm in the next section and use it to formulate the definition of D-reducibility. In the section after next, subsequent to the treatment of D-reducibility, we will examine A-, B-, and C-reducibility for the first time. This is in reverse order to the historical development.

6.4   The Dürre–Heesch Algorithm

Suppose we are given a configuration $C = (E_C, \mathcal{L}_C)$. Denote by R the ring of outer vertices and by r the ring size of C. First we consider how the outer vertices can be colored. From now on, we will be using only the numbers 0, 1, 2, and 3 as colors—as was the case in the proof of the reducibility of the Birkhoff diamond (page 181).

Boundary Colorings

The vertices of R will be denoted by the sequence $\mathbf{x}_1, \ldots, \mathbf{x}_r$ ordered cyclically. A coloring of the vertices $\mathbf{x}_j$ for $j \in \{1, \ldots, r\}$ can be thought of as an r-tuple of colors and hence as an element of the set $\{0, 1, 2, 3\}^r$. Naturally, not all r-tuples yield admissible colorings. In order to obtain an overview, we will now proceed to formulate these ideas abstractly. Addition and subtraction *modulo r* are useful in this regard—meaning that we add and subtract as usual up to $r + 1 = 1$ and $1 - 1 = r$.

Definition 6.4.1
(a) Let r be a natural number such that $r \geq 3$. An r-tuple $\boldsymbol{a} = (a_1, \ldots, a_r) \in \{0, 1, 2, 3\}^r$ is said to be a *boundary coloring (of size r)* if $a_j \neq a_{j+1}$ for all $j \in \{1, \ldots, r\}$.
(b) Two boundary colorings $\boldsymbol{a}$ and $\boldsymbol{a}'$ are said to be *equivalent* if they have the same size r and differ only by a permutation of colors. This means that there must exist a bijection $\pi : \{0, 1, 2, 3\} \to \{0, 1, 2, 3\}$ such that $a_j' = \pi(a_j)$ for all $j \in \{1, \ldots, r\}$ (compare the definition of equivalence on page 86).

The set of all boundary colorings of size r will be denoted by $\Phi(r)$. Equivalence of boundary colorings is obviously an equivalence relation. Frequently, it suffices to consider a particular representative from each equivalence class. In order to obtain such a representative, the boundary colorings of equal size will be ordered "lexicographically." By this we mean the linear ordering defined as follows: $\boldsymbol{a} < \boldsymbol{a}'$ if there exists an index j_0 such that $a_j = a_j'$ for $j < j_0$ and $a_{j_0} < a_{j_0}'$. The designation of a boundary coloring within an equivalence class is now easily achieved.

Definition 6.4.2
A boundary coloring is said to be *essential* if it is the smallest, with respect to the lexicographic ordering, within its equivalence class.

The following must hold for an essential boundary coloring a: $a_1 = 0$, $a_2 = 1$, $a_3 \in \{0, 2\}$. For boundary colorings of even size r, there exists (exactly) one essential coloring having only two colors, namely $(0, 1, 0, 1, \ldots, 0, 1)$. For boundary colorings of odd size r, at least three colors will always be required. In the discussion of the Birkhoff diamond on page 182, the second, fourth, and sixth boundary colorings are essential. The first is equivalent to the essential boundary coloring $(0, 1, 0, 1, 2, 3)$, the third to $(0, 1, 2, 1, 2, 3)$, and the fifth to $(0, 1, 2, 1, 0, 3)$. The total number $f(r)$ of essential boundary colorings of size r can be easily calculated.

Lemma 6.4.3

$$f(r) = \begin{cases} \frac{3^{r-1} - 1}{8} & \text{for } r \text{ odd,} \\[2ex] \frac{3^{r-1} + 5}{8} & \text{for } r \text{ even.} \end{cases} \qquad \blacksquare$$

Proof If $r = 3$, we have only one essential boundary coloring, namely, $(0, 1, 2)$. For $r = 4$, we find four essential boundary colorings:

$$(0, 1, 0, 1), \quad (0, 1, 0, 2), \quad (0, 1, 2, 1), \quad (0, 1, 2, 3).$$

Hence, the function f has the appropriate values for $r = 3$ and $r = 4$.

We proceed by induction and choose an $r > 4$ with the assumption that $f(r')$ for all $r' < r$ has the form given in the statement of the lemma. For an essential boundary coloring $a = (a_1, \ldots, a_{r-1}, a_r)$, we consider the $(r-1)$-tuple $a^\flat = (a_1, \ldots, a_{r-1})$ and the $(r-2)$-tuple $a^{\flat\flat} = (a_1, \ldots, a_{r-2})$. If $a^\flat$ is a boundary coloring of size $r - 1$, then we say that a is a *boundary coloring of type* 1. This happens precisely when $a_{r-1} \neq 0$. In this case, $a^\flat$ is also essential. If $a_{r-1} = 0$, then we say that a is a *boundary coloring of type* 2. Then $a^{\flat\flat}$ is an essential boundary coloring of size $r - 2$. It is evident that each boundary coloring of size r is either of type 1 or of type 2, but not both. There are two cases to consider:

r **odd**: If *r* is odd, then $r - 1$ is even. Then the case $a^b = (0, 1, 0, 1, \ldots, 0, 1)$ arises. It can be extended in only one way to an essential boundary coloring of size *r*, namely, to $a = (0, 1, 0, 1, \ldots, 0, 1, 2)$. The reason for this is the following: If $a_r = 0$ or 1, no boundary colorings are produced, and if $a_r = 3$, one obtains a boundary coloring that is not essential. All other boundary colorings of size $r - 1$, however, can be extended to a boundary coloring of size *r* in exactly two ways, namely, when $a_r \neq 0, a_{r-1}$. In this way, all boundary colorings of type 1 can be realized. By the induction hypothesis, it follows that the number of such colorings is

$$2 \cdot \frac{3^{r-2} + 5}{8} - 1 = \frac{2 \cdot 3^{r-2} + 2}{8}.$$

The boundary colorings *a* of type 2 can be reclaimed from the boundary colorings a^{bb} of size $r - 2$ by setting $a_{r-1} = 0$ and letting $a_r \in \{1, 2, 3\}$. Since $r - 2$ is again odd, in each a^{bb}, the colors 0, 1, and 2 already appear. Therefore, there are always these three possibilities for a_r. By the induction hypothesis, the number of boundary colorings of type 2 is

$$3 \cdot \frac{3^{r-3} - 1}{8} = \frac{3^{r-2} - 3}{8}.$$

By adding, one now obtains the desired value

$$f(r) = \frac{3^{r-1} - 1}{8}.$$

r **even**: Now $r - 1$ is odd. Every boundary coloring of size $r - 1$ contains at least the three colors 0, 1, 2 and can be extended to a boundary coloring of size *r* in precisely two ways. Therefore, the number of boundary colorings of type 1 is

$$2 \cdot \frac{3^{r-2} - 1}{8} = \frac{2 \cdot 3^{r-2} - 2}{8}.$$

To determine the boundary colorings of type 2, one observes that the boundary coloring $(0, 1, 0, 1, \ldots, 0, 1)$ of size $r - 2$ can be extended to an essential boundary coloring of type 2 in only two ways, namely to $(0, 1, 0, 1, \ldots, 0, 1, 0, 1)$ and to $(0, 1, 0, 1, \ldots, 0, 1, 0, 2)$. All other boundary colorings of size $r - 2$, on the other hand, have three possible extensions. By the induction hypothesis, the number of boundary colorings of type 2 is

$$3 \cdot \frac{3^{r-3} + 5}{8} - 1 = \frac{3^{r-2} + 7}{8}.$$

In this case as well, the desired value

$$f(r) = \frac{3^{r-1} + 5}{8}$$

is achieved.   □

In his first computer program, mainly to save on memory requirements, Dürre had assigned to each boundary coloring a unique natural number in which the individual entries of the tuple were interpreted as specific digits of a 4-adic number. The *Dürre number* F of a boundary coloring a is defined to be

$$F = \sum_{j=1}^{r} a_j 4^{r+1-j}.$$

Later on, it turned out that much more important than the Dürre number were the algorithms to determine the ordinal of a boundary coloring from its lexicographical order and, conversely, to recover the boundary coloring from its ordinal. These algorithms were also developed by Dürre. They made it possible to reduce considerably the amount of memory required. It became the major breakthrough in the handling of larger, more complex diagrams.

A boundary coloring of a given size is obviously uniquely determined by its Dürre number. Therefore, one can also order the boundary colorings by the size of their Dürre numbers. That, however, yields nothing new. It produces the lexicographical ordering back again.

In a slight modification of the terminology, we speak of a *boundary coloring of the configuration C* if we mean a vertex 4-coloring of the bounding circuit of C. After establishing a numbering of the outer vertices of C in cyclic order, we can identify the boundary colorings of C with (abstract) boundary colorings whose size coincides with the ring size of C. In doing this, we interpret a component of a boundary coloring as the color of the vertex with the same number.

TABLE 6.1 The Birkhoff diamond essential boundary colorings of size 6

Running Number	Boundary Coloring	Dürre Number	Running Number	Boundary Coloring	Dürre Number
1	(0,1,0,1,0,1)	1092	2	(0,1,0,1,0,2)	1096
3	(0,1,0,1,2,1)	1124	4	(0,1,0,1,2,3)	1132
5	(0,1,0,2,0,1)	1156	6	(0,1,0,2,0,2)	1160
7	(0,1,0,2,0,3)	1164	8	(0,1,0,2,1,2)	1176
9	(0,1,0,2,1,3)	1180	10	(0,1,0,2,3,1)	1204
11	(0,1,0,2,3,2)	1208	12	(0,1,2,0,1,2)	1560
13	(0,1,2,0,1,3)	1564	14	(0,1,2,0,2,1)	1572
15	(0,1,2,0,2,3)	1580	16	(0,1,2,0,3,1)	1588
17	(0,1,2,0,3,2)	1592	18	(0,1,2,1,0,1)	1604
19	(0,1,2,1,0,2)	1608	20	(0,1,2,1,0,3)	1612
21	(0,1,2,1,2,1)	1636	22	(0,1,2,1,2,3)	1644
23	(0,1,2,1,3,1)	1652	24	(0,1,2,1,3,2)	1656
25	(0,1,2,3,0,1)	1732	26	(0,1,2,3,0,2)	1736
27	(0,1,2,3,0,3)	1740	28	(0,1,2,3,1,2)	1752
29	(0,1,2,3,1,3)	1756	30	(0,1,2,3,2,1)	1764
31	(0,1,2,3,2,3)	1772			

In the case of the Birkhoff diamond, we have ring size 6, and therefore we have to consider 31 essential boundary colorings (see Table 6.1). [2]

Color-Extendibility

The first step towards *D*-reducibility consists in examining the essential boundary colorings of the configuration for "color-extendibility."

---

[2]This table was created using Dürre's program. It establishes that a boundary coloring is generated by its running number (that is, its lexicographical ordinal).

This means establishing which of the essential boundary colorings can be extended to the entire configuration without being altered in any way. Dürre called these boundary colorings *directly color-extendible*; Appel and Haken designated them as *good from the outset*. One should observe that the set $\Phi(r)$ of all boundary colorings of size r depends only upon the ring size r of the configuration C under consideration. On the other hand, the set of boundary colorings that are good from the outset, which we will denote by $\Phi_0(C)$, is determined by the entire structure of C. It is clear that each boundary coloring equivalent to a coloring that is good from the outset is again good from the outset. This means that the set $\Phi_0(C)$ is *closed* with respect to equivalence.

For the practical execution of the task at hand, one must first of all program the information describing the entire configuration into the computer. This takes the following format. Let w inner vertices be given; denote them by $y_1, \ldots, y_w$. For the mechanical processing, all vertices will be renumbered, that is, we will set

$$z_j = \begin{cases} x_j & \text{for } 1 \leq j \leq r, \\ y_{j-r} & \text{for } r < j \leq r + w. \end{cases}$$

The configuration is then completely described arithmetically through the numbers β_{jk} for $1 \leq j < k \leq r + w$, which are defined as follows:

$$\beta_{jk} = \begin{cases} 1 & \text{if } z_j \text{ and } z_k \text{ are joined by an edge in } C, \\ 0 & \text{otherwise.} \end{cases}$$

In particular, for the indices whose corresponding vertices belong to the ring R, it follows that

$$\beta_{j\,j+1} = 1 \quad \text{for } 1 \leq j < r,$$
$$\beta_{jk} = 0 \quad \text{for } 1 \leq j < k \leq r,\, k \neq j+1,$$
$$\beta_{1r} = 1.$$

The numbers β_{jk} can then be inserted into what is termed an *adjacency matrix*. This matrix, which completely describes the configuration in a numerical format, can be programmed into the computer. The adjacency matrix of the Birkhoff diamond is shown below.

ADJACENCY MATRIX

	1	2	3	4	5	6	7	8	9	10
1	0	1	0	0	0	1	1	1	0	0
2	1	0	1	0	0	0	0	1	1	0
3	0	1	0	1	0	0	0	0	1	0
4	0	0	1	0	1	0	0	0	1	1
5	0	0	0	1	0	1	1	0	0	1
6	1	0	0	0	1	0	1	0	0	0
7	1	0	0	0	1	1	0	1	0	1
8	1	1	0	0	0	0	1	0	1	1
9	0	1	1	1	0	0	0	1	0	1
10	0	0	0	1	1	0	1	1	1	0

To prove direct color-extendibility, Dürre formed *color matrices* consisting of $4 \times (r + w)$ matrices (v_{ij}), $v_{ij} \in \{0, 1\}$, in which each column contained exactly one 1 and such that for all $i \in \{0, 1, 2, 3\}$ and $j, k \in \{1, \ldots, r + w\}$,

$$v_{ij} = v_{ik} = 1 \implies \beta_{jk} = 0.$$

The four rows of such a matrix correspond to the four colors and therefore will be indicated by the numbers $0, 1, 2, 3$. The columns will denote the vertices of the configuration. If $a_{ij} = 1$, then a vertex j has the color i. The last stated condition on the color matrix implies that vertices with the same color are not neighbors. Hence, the coloring is admissible. A boundary coloring establishes the entries in the first r columns of the color matrix, and direct color-extendibility occurs if a complete color matrix can be formed from them.

In the case of the Birkhoff diamond, the sixteen boundary colorings with the numbers 3, 4, 5, 6, 8, 11, 14, 19, 20, 21, 22, 24, 25, 26, 27, and 30 turn out to be directly color-extendible. For instance, the color matrix of boundary coloring number 24 (that is, the second coloring of outer vertices on page 182) is given below.

$$\begin{pmatrix} 1 & 0 & 0 & 0 & 0 & 0 & 0 & 0 & 0 & 1 \\ 0 & 1 & 0 & 1 & 0 & 0 & 1 & 0 & 0 & 0 \\ 0 & 0 & 1 & 0 & 0 & 1 & 0 & 1 & 0 & 0 \\ 0 & 0 & 0 & 0 & 1 & 0 & 0 & 0 & 1 & 0 \end{pmatrix}.$$

Furthermore, boundary colorings 4, 19, 20, and 22 (see Table 6.1) also cropped up in our earlier treatment of the Birkhoff diamond.

Chromodendra

Before we are able to continue the description of the Dürre–Heesch algorithm, we require a little more theory.

Let the configuration C be embedded in the minimal triangulation G. In the context of D-reducibility, a *proper* embedding is not absolutely necessary (Definition 5.4.9). If we remove from G the inner vertices of C and the edges incident with them, then we obtain a graph G' which we can, by assumption, color with four colors. Every coloring of G' induces a boundary coloring of C.

In this subsection, the core of C is no longer of the essence. Therefore, we will be mainly considering colored connected graphs G' without bridges and final edges all of whose faces, with exactly one exception, are bordered by triangles. The vertices and edges of the *exceptional face* form the *bounding circuit* of G'. If $x_1, \ldots, x_r$ are the vertices of the bounding circuit labeled in cyclic order, then in this case as well, we can identify the boundary colorings of size r with the vertex 4-colorings of the bounding circuit (see page 190). Now, every edge belongs to the boundary of exactly two faces (Theorem 2.6.8). In particular, each edge that does not belong to the bounding circuit belongs to the boundary of exactly two triangles.

By a *color-pair choice* we mean the choice of a color $w \in \{1, 2, 3\}$ that results in a partition of the set of colors into the two color-pairs $\{0, w\}$ and $\{1, 2, 3\} \setminus \{w\}$. This also produces Kempe nets in G' and in the bounding circuit of G'. The Kempe nets in the bounding circuit are called *Kempe sectors,* in accordance with [DÜRRE and MIEHE 1979]. They have very simple structures. The set of vertices of a Kempe sector can always be arranged to form a chain. The number of Kempe sectors is either 1 or is an even number. There are two types of sectors: *w-sectors*, whose vertices are colored with 0 or w, and $\overline{w}$-*sectors*, having vertices colored with the two remaining colors. Different Kempe sectors of the same type can belong to the same Kempe net in G'. We call a set of vertices of the bounding circuit a *block* if the vertices belong to the same Kempe net of G' and if none of the remaining vertices of this Kempe net lie in the bounding circuit. Thus we have a decomposition of the set of vertices of the bounding circuit into blocks. Such a decomposition is said to be a *block decomposition with respect to the given boundary coloring and the color-pair choice*. Each block is a union of the vertex sets of a number of Kempe sectors of the same type. Accordingly, we have two types of blocks: w-blocks and $\overline{w}$-blocks.

The corresponding decompositions of the (abstract) boundary colorings into blocks is important for the Kempe chain games. One requires an overview of all possible block decompositions of a boundary coloring. Each block decomposition offers the possibility of one or more Kempe interchanges. By considering a boundary coloring of a configuration, one can then hope to obtain a "better" boundary coloring by such a Kempe interchange. What is interesting and important is that the possible block decompositions are determined solely by the boundary coloring and the color-pair choice. They are independent of the configurations and the graphs in which the configurations are embedded.

In order to make this somewhat clearer, we must explicitly describe block decompositions of abstract boundary colorings. We will use the following terminology. Let r be a natural number. A *decomposition* of the set $M = \{1, \ldots, r\}$ is a family $B_1, \ldots, B_s$ of nonempty pairwise disjoint subsets of M that completely *cover* M; that is, whose union is equal to M. The individual members of such a family we again designate as blocks. We say that two distinct blocks B_k and B_l *abut each other* at the point t if one of the numbers $t, t+1$ belongs to B_k and the other to B_l. Finally, we carry over the notion of a Kempe sector to abstract boundary colorings. A *Kempe sector of a boundary coloring* $\boldsymbol{a} = (a_1, \ldots, a_r)$ with respect to the color-pair choice w is a sequence $(t_1, \ldots, t_p)$ of indices such that the following conditions hold:[3]

1. $t_{i+1} = t_i + 1$ for all $i \in \{1, \ldots, p-1\}$.

2. All a_{t_i} are elements of the same color-pair.

---

[3]Here again we are using addition and subtraction modulo r. See page 187.

3. a_{t_1-1} and a_{t_p+1} are elements of color-pairs that are complementary to the pairs in part 2.

In the event that the color-pair in part 2 is the pair $\{0, w\}$, then we call it, in particular, a w-sector. Otherwise, we have $\overline{w}$-sectors. The maximal elements of the individual Kempe sectors are called *points of contact*. At these points the Kempe sectors of different types *abut* one another. If t is a point of contact, then t and $t + 1$ belong to different Kempe sectors. For a fixed boundary coloring and color-pair choice, the number of Kempe sectors is either 1 or is an even number. The number of points of contact is therefore always even. Hence, we are able to formulate and prove the following result.

Theorem 6.4.4

Let G' be a colored connected graph without bridges or final edges all of whose faces, other than its unique exceptional face, are bordered by triangles. Suppose a boundary coloring $\boldsymbol{a} = (a_1, \ldots, a_r)$ of size r and a color-pair choice $w \in \{1, 2, 3\}$ are given. Then a partition of the index set $\{1, \ldots, r\}$ into blocks $B_1, \ldots, B_s$ is a block decomposition (with respect to $\boldsymbol{a}$ and w) if and only if:

1. *Each block is a union of Kempe sectors of the same type.*
   Notation: w-block, respectively $\overline{w}$-block.
2. *Blocks cannot mutually overlap. In other words, for $k_1, k_2 \in B_k$ and $l_1, l_2 \in B_l$, $k \neq l$, the order*

$$k_1 < l_1 < k_2 < l_2$$

   is not possible.
3. *If two blocks abut at one point, then they abut at exactly one more point.* ■

Proof First of all, we show that the given conditions are necessary. The first condition follows immediately from the definition. The second condition we have already proven earlier (Lemma 6.1.5).

For the proof of the third condition, we first note that because of condition 2, two blocks can abut at a maximum of two points. Now suppose that the Kempe nets G_1 and G_2 in G' (which arose as described earlier on page 194) induce on its bounding circuit the blocks B_1 and B_2, which abut at the vertex $\mathbf{x}_1$. We are further able to assume that B_1 is a w-block that contains $\mathbf{x}_1$ as a vertex. Hence, $\mathbf{x}_1$ is also a vertex of G_1. Then B_2 is a $\overline{w}$-block that contains, as does

G_2, the vertex x_2. The vertices x_1 and x_2 are neighbors in G' and therefore are joined by an edge K_2. As every edge in G' belongs to the boundary of a triangle, we can find a vertex z_3 that extends the pair $\{x_1, x_2\}$ to a triangle. From the definition of Kempe net, the vertex z_3 must belong either to G_1 or to G_2 regardless of its color—that is, it may be colored with 0, with w, or with one of the other two colors. We let i_3 denote the index for which $z_3 \in G_{i_3}$ holds and note that the other possible index can be represented in the form $3 - i_3$. If the edge K_3 that joins z_3 with x_{3-i_3} belongs to the corresponding bounding circuit, then z_3 is either x_3 or x_r. Hence, the two blocks also abut at either the point 2 or at the point r. (This can, of course, happen only when the embedding is not a proper embedding.) On the other hand, if K_3 does not belong to the bounding circuit, then K_3 belongs to the boundary of a second triangle, whose third vertex we will denote by z_4. This vertex z_4 again belongs to a G_{i_4}, where $i_4 \in \{1, 2\}$, and we have an edge K_4 that joins z_4 with the vertex of K_3 belonging to G_{3-i_4}. If K_4 belongs to the bounding circuit, then we are finished. Otherwise, we can find a vertex z_5 and edge K_5 with the appropriate properties. We can continue this process until we obtain an edge K_n belonging to the bounding circuit. As G' is finite, this must occur at some time or other. The edge K_n thus obtained is certainly distinct from K_2. If x_t and x_{t+1} are the end points of K_n (addition modulo r), then the blocks B_1 and B_2 now also abut one another at the point $t \neq 1$.

Conversely, for a given partition of the index set $\{1, \ldots, r\}$ satisfying the given conditions 1 to 3, we must construct a colored graph G' that has the given properties and on whose bounding circuit the desired block decomposition occurs. We note that from the original basic approach, the exceptional face of such a graph is a bounded face. It is the interior domain of a configuration. However, it is also true that this face, by a double application of the stereographic projection, can be transformed into the unbounded face without having lost its defining properties. For this reason, it suffices to construct G' so that the exceptional face is unbounded.

We begin with a regular r-gon whose vertices $x_1, \ldots, x_r$ are given in cyclic order. These vertices and the sides of the r-gon (treated as edges) form the bounding circuit of the desired graph G'. The vertices will be colored in accordance with the given boundary coloring a.

It can happen that only one block exists. Then only two colors will be needed. The vertices x_j can be colored alternately with two colors. In particular, it follows that r is an even number. We add the center of the regular r-gon, which is colored with a third color, as a vertex, and we add, as edges, the line segments that join the center to the vertices x_j. In this way, the desired graph G' is constructed.

Now we assume that we have several blocks. Then there exist points at which the blocks abut one another. The number of points of contact is, by property 3, even. Suppose it is $2m$ for some natural number m. Denote the points of contact by $t_1, \ldots, t_m, u_1, \ldots, u_m$. Then the following conditions hold:

1. $1 \leq t_1 < t_2 < \cdots < t_m \leq r$.

2. $t_i < u_i$ for all $i \in \{1, \ldots, m\}$.

3. x_{t_i} and x_{u_i+1} belong to the same block for all $i \in \{1, \ldots, m\}$.

Then the vertices x_{t_i+1} and x_{u_i} belong to the same block again for every $i \in \{1, \ldots, m\}$. Throughout all this, it is possible that $u_i = t_i + 1$. This is precisely the case if u_i forms a block in its own right.

Now, for all $i \in \{1, \ldots, m\}$, we add as edges the line segments from x_{t_i} to x_{u_i+1}, as well as, for i such that $u_i > t_i + 1$, the line segments from x_{t_i+1} to x_{u_i}. From condition 2, it follows that these new edges do not intersect one another. At most, they can abut at an end point. Of course, the vertex coloring may now no longer be admissible. Both end points of one of the newly added edges might be colored with the same color. However, that is easily fixed. Such an edge can be subdivided through its midpoint. Two edges and one additional vertex result from this. The vertex can take on the second color of the color-pair to which the color of the original vertices of the subdivided edge belongs. The regular r-gon is by this time subdivided into convex polygons of two types. The first type consists of polygons whose vertices are colored alternately with the two colors of the same color-pair. The second type consists of polygons with at most 6 vertices such that the induced boundary coloring decomposes into exactly two Kempe sectors of which each contains at most three elements. To each polygon of the first type, we choose an interior point as a new vertex, color it with a color of the complementary color-pair, and add as edges the line segments joining this new vertex to the original vertices of the polygon. In doing this, we have subdivided these polygons into triangles. For the polygons of the second type, we need make alterations only if there is a polygon with at least four vertices. In this case, we find a vertex whose color has appeared only once in this polygon. We add as further edges the diagonals emanating from this vertex. In this way, the coloring remains admissible, and the entire regular r-gon is subdivided into triangles, as was desired. □

We now approach the subject of a chromodendron, which is the central character in this section. Suppose a boundary coloring a, a color-pair choice w, and the corresponding block decomposition are all given. The corresponding *chromodendron* is the combinatorial graph whose vertices are the blocks and whose edges are pairs of abutting blocks. This choice of terminology [TUTTE and WHITNEY 1972], whose etymology goes back to ancient Greek, is made clear from the following result.

Theorem 6.4.5
A chromodendron is a tree. ∎

Proof Since a circuit is connected, one can get from each block to any other block by a sequence of abutting blocks. Hence, a chromodendron is connected.

What remains to be shown is that a chromodendron is circuit-free. That will be established geometrically. We consider the given boundary coloring of size r as the coloring of a circuit with r vertices. Then we have a partition of this circuit into Kempe sectors and a block decomposition of the set of vertices. If, in the circuit, we remove the vertices of a block that abuts at least two other blocks together with the edges incident with them, then there are distinct components remaining. The vertices of any other block must, by condition 2 of the previous theorem, lie entirely in one of the components. Therefore, two blocks lying in different components cannot be linked by a sequence of abutting blocks. This means that after the removal of this block, the chromodendron splits apart. Hence, no circuit can exist in the chromodendron. □

We take note of a simple consequence of the foregoing deliberations.

Corollary 6.4.6
In a fixed boundary coloring and for a fixed color-pair choice, the number of blocks in a block decomposition is either 1 or is 1 plus one-half the number of Kempe sectors. ■

Proof We can assume that more than one Kempe sector exists. Hence, the number q of Kempe sectors is even. Then two Kempe sectors abut each other if and only if two blocks also abut each other. For each pair of abutting blocks there are exactly two points of contact. Thus the corresponding chromodendron has $\frac{q}{2}$ edges. Because in a tree we are dealing with a connected graph with only one face, it follows from the Euler polyhedral formula (Theorem 4.3.3) that the number of vertices of a tree is equal to the number of its edges plus 1. □

The Kempe Chain Game

At this point, we go back to the presentation of the Dürre–Heesch algorithm. A configuration C is certainly reducible if each boundary coloring is good from the outset. This means that

$$\Phi_0(C) = \Phi(r).$$

Unfortunately, this is seldom the case. Roughly speaking, D-reducibility occurs when each boundary coloring can be transformed by a number of Kempe interchanges into a coloring that is good from the outset. The following parts of the Dürre–Heesch algorithm verify this.

We will explain the next steps using the example of the Birkhoff diamond. The calculations thus far have shown that boundary coloring number 1, namely, $(0, 1, 0, 1, 0, 1)$, is not directly color-extendible. We now examine which Kempe interchanges are possible.

The color-pair choice $w = 1$ yields nothing. We obtain only one Kempe sector. The only possible Kempe interchange yields the nonessential boundary coloring $(1, 0, 1, 0, 1, 0)$. The corresponding

TABLE 6.2 Block decompositions of the Birkhoff diamond boundary coloring number 1 of size 6 for the color-pair choice $w = 2$

No.	B_1	B_2	B_3	B_4
1.	$\{1\}$	$\{3\}$	$\{5\}$	$\{2, 4, 6\}$
2.	$\{2\}$	$\{4\}$	$\{6\}$	$\{1, 3, 5\}$
3.	$\{1\}$	$\{4\}$	$\{2, 6\}$	$\{3, 5\}$
4.	$\{2\}$	$\{5\}$	$\{1, 3\}$	$\{4, 6\}$
5.	$\{3\}$	$\{6\}$	$\{1, 5\}$	$\{2, 4\}$

essential boundary coloring is the boundary coloring with which we started.

From the structure of boundary coloring 1, it is clear that the color-pair choices 2 and 3 must be equivalent. The computer does the calculations for the next possibility, that is, when $w = 2$. Now, each of the Kempe sectors contains a single index. We have a total of six Kempe sectors and therefore block decompositions each consisting of four blocks.[4]

There exist five possible block decompositions (see Table 6.2).

Each of these block decompositions can be cracked open externally. Therefore, it is now a matter of proving whether in each case it is possible to apply a Kempe interchange leading to a boundary coloring that is good from the outset. We play the game step by step from one block decomposition to the next.

1. In block decomposition 1, the first possibility for a Kempe interchange consists in replacing color 0 by color 2 in B_1. This yields the boundary coloring $(2, 1, 0, 1, 0, 1)$. The corresponding essential boundary coloring is $(0, 1, 2, 1, 2, 1)$, which is boundary coloring

---

[4]At this point, it should be noted that Heesch suspected that the greatest difficulties would occur in the mechanical generation of block decompositions [BIGALKE 1988, page 175]. However, it turned out that the actual difficulties, whose solution formed the basis for Dürre's doctoral thesis, were caused by the generation and coding of the boundary colorings. See also footnote 2.

number 21. It is good from the outset. This is nice indeed. We need not examine this block decomposition any further.

2. Analogously, for block decomposition 2, we replace color 1 with color 3 in B_1. We thus obtain the boundary coloring $(0, 3, 0, 1, 0, 1)$, which is equivalent to the essential boundary coloring $(0, 1, 0, 2, 0, 2)$. The latter is boundary coloring number 6, and it is also good from the outset.

3. We proceed as in step 1 for block decomposition 3 and obtain the same nice result.

4. We proceed as in step 2 for block decomposition 4 and obtain the same nice result.

5. In block decomposition 5, we replace color 0 by color 2 in B_1 and thereby immediately obtain an essential boundary coloring, namely $(0, 1, 2, 1, 0, 1)$. However, this is boundary coloring number 18, which is not directly color-extendible. This is not as nice, but we are still not forced into giving up. Let us try again. Instead, we replace color 1 by color 3 in B_2. This leads to $(0, 1, 0, 1, 0, 3)$. The corresponding essential boundary coloring is $(0, 1, 0, 1, 0, 2)$. This is boundary coloring number 2 and is also not directly color-extendible. So now what? Well, we recolor B_1 and B_2 simultaneously. This immediately yields the essential boundary coloring $(0, 1, 2, 1, 0, 3)$, which is number 20 and which is good from the outset.

Thus, in all five cases, we have achieved the desired result. We say that boundary coloring number 1 of the Birkhoff diamond is good from stage 1. In general:

Definition 6.4.7

A boundary coloring of the configuration C is *good of stage* 1, or *class 1 good*, if it is not good from the outset but, after a choice of a color-pair, each block decomposition permits a Kempe interchange transforming it into a boundary coloring that is good from the outset, that is, into an element of $\Phi_0(C)$.[5]

---

[5]This is a horrible definition, with three quantifiers: $\exists$ color-pair choice $\forall$ block decomposition $\exists$ Kempe interchange . . . . However, local continuity of a real function also requires three quantifiers, and global continuity has even four!

In the case of the Birkhoff diamond, other boundary colorings that turn out to be good from stage 1 are numbers 2, 10, 18, and 31.

We denote by $\Phi_1(C)$ the set of boundary colorings consisting of $\Phi_0(C)$ together with the boundary colorings that are good from stage 1. What we have just shown in the foregoing example is that the Dürre–Heesch algorithm contains a mechanical procedure for establishing class 1 goodness. In order to continue with the discussion of the Dürre–Heesch algorithm, it is advisable to generalize the definition of goodness of a boundary coloring just a little bit more. To do this, we must first define the moves in the Kempe chain game.

Definition 6.4.8
Let a be a boundary coloring of size r, w a color-pair choice, and $B_1, \ldots, B_s$ a corresponding block decomposition. The boundary coloring a is said to be *transformed* by a *Kempe interchange*, or by a *recoloring*, into the boundary coloring b if b results from a when the colors 0 and w in certain w-blocks and/or both of the other colors in certain $\overline{w}$-blocks are interchanged.

For a given boundary coloring, color-pair choice, and block decomposition, a Kempe interchange—hence, a move in the Kempe chain game—is determined by the representation of the blocks in which the interchange takes place. Thus the number of possible moves is finite and is equal to the number of *block selections*. However, different selections can lead to equivalent results. For instance, if one performs an interchange within all w-blocks, then one obtains a coloring that is equivalent to the original coloring. The same thing happens if one performs an interchange within all $\overline{w}$-blocks or overall throughout all blocks.

Definition 6.4.9
Let Φ be a set of boundary colorings of size r. A boundary coloring $a \in \Phi(r)$ is said to be Φ-*good* if it does not belong to Φ itself but if there exists a color-pair choice such that for each corresponding block decomposition, a Kempe interchange exists that transforms a into an element of Φ.

Since a Kempe interchange, in general, does not preserve the property of being essential in a boundary coloring, for all practical purposes it is sensible to assume that the set Φ of boundary colorings mentioned in this definition is closed with respect to equivalence. Then, however, Φ-goodness is also closed with respect to equivalence.

Lemma 6.4.10
Let Φ be a set of boundary colorings of fixed size that is closed with respect to equivalence. Then each boundary coloring that is equivalent to a Φ-good boundary coloring is also Φ-good.   ∎

Proof Let $a = (a_1, \ldots, a_r)$ be a Φ-good boundary coloring, and let φ be a permutation of colors. We must show that $a' = (a'_1, \ldots, a'_r)$, where $a'_j = \varphi(a_j)$ for all $j \in \{1, \ldots, r\}$, is Φ-good as well. Since by assumption Φ is closed with respect to equivalence, we have in any case that $a' \notin \Phi$.

Furthermore, let w be a color-pair choice such that for each corresponding block decomposition, a Kempe interchange exists that transforms a into an element of Φ. Let the color-pair choice w' be determined by the condition that the pair $\{\varphi(0), \varphi(w)\}$ is a corresponding color-pair, and let $B'_1, \ldots, B'_s$ be a block decomposition with respect to a' and w'. By letting the vertices in each block be represented by their colors and not their indices, the mapping φ can be extended to a mapping of blocks. Then $B_1, \ldots, B_s$ where $B_k = \varphi^{-1}(B'_k)$ for all $k \in \{1, \ldots, s\}$, is a block decomposition corresponding to w with respect to a and w. Now, if $\{B_{k_1}, \ldots, B_{k_t}\}$ is a selection of blocks such that an appropriate Kempe interchange applied to these blocks transforms a into a boundary coloring $b \in \Phi$, then the Kempe interchange corresponding to the block selection $\{B'_{k_1}, \ldots, B'_{k_t}\}$ transforms a' into a boundary coloring that is equivalent to b through φ and that also belongs to Φ, because Φ is closed with respect to equivalence.   □

A consequence of this fact is that the portion of the Dürre–Heesch algorithm that has been developed up to now provides a procedure for determining the property of Φ-goodness where Φ is a specified set of boundary colorings of a fixed size that is closed with respect to equivalence.

If a configuration C of ring size r is given, then the set $\Phi_0(C)$ of boundary colorings of C that are good from the outset is determined. The boundary colorings of C of class 1 goodness, which were defined earlier, coincide exactly with the $\Phi_0(C)$-good boundary colorings. The remainder of the Dürre–Heesch algorithm consists in a repetitive process that determines higher orders of Φ-goodness. From this we obtain higher classes of goodness that are determined inductively.

Definition 6.4.11
Let C be a configuration, n a natural number, and $\Phi_n(C)$ the set of already established boundary colorings of goodness class less than or equal to n. A boundary coloring is said to be *good of stage $n + 1$* if it is $\Phi_n(C)$-good.

For a given configuration C of ring size r, an increasing nested chain of sets of boundary colorings $\Phi_n(C)$ is produced. Since the set $\Phi(r)$ is finite, this chain must at some time or other become stationary. Thus there must exist an index n_0 such that no boundary colorings of goodness class $n_0 + 1$ exist. Because it does not depend on the specific value of n_0, we write (as Heesch did) $\overline{\Phi}(C)$ instead of $\Phi_{n_0}(C)$. There are two distinct cases to consider. These we will cover in the following definition—a definition that has already been long expected.

Definition 6.4.12
1. The configuration C is said to be reducible—more precisely, it is said to be *D-reducible*—if $\overline{\Phi}(C) = \Phi(r)$, meaning that each of its boundary colorings is good of some stage or other.
2. The configuration C is said to be *D-irreducible* if $\overline{\Phi}(C)$ is a proper subset of $\Phi(r)$.

In the case of the Birkhoff diamond, $\overline{\Phi} = \Phi_5 = \Phi(r)$:

Stage	Boundary Coloring No.
2	7, 23
3	9, 15, 16, 29
4	13, 24, 28
5	17

In order to include some historical details, we will show the essential part of the recoloring procedure from Dürre's first computer program, which was run on November 23, 1965.

```
≠PROCEDURE≠ UMFARB(N,U,B,KOMPL,KET,DB,DFB).,
≠COMMENT≠ UMFAERBUNG.,
≠INTEGER≠ N,U.,    ≠INTEGER≠≠ARRAY≠ B,KOMPL, KET,DB,DFB.,
≠BEGIN≠
  ≠INTEGER≠ I,J,K,ANZ,D.,≠INTEGER≠≠ARRAY≠ NORM(/1.,2/).,
  ANZ.=(U+1)*2.,
  ≠FOR≠ I.=1 ≠STEP≠ 1 ≠UNTIL≠ U ≠DO≠
  ≠BEGIN≠
    ≠IF≠ DFB(/I/) ≠EQUAL≠ 0 ≠THEN≠ ≠GO TO≠ AUSI.,
    K.=DB(/DFB(/I/),1/).,
      M11..
    ≠IF≠ K ≠EQUAL≠ 1 ≠THEN≠
    ≠BEGIN≠
      ≠IF≠ KET(/ANZ+1/) ≠NOT EQUAL≠ 0 ≠THEN≠ K.=ANZ+1.,
    ≠END≠.,
    M2.,
    ≠FOR≠ J.=KET(/K-1/)+1 ≠STEP≠ 1 ≠UNTIL≠ KET(/K/) ≠DO≠
        B(/J/).=KOMPL(/B(/J/)/).,
    ≠IF≠ K ≠EQUAL≠ ANZ+1 ≠THEN≠ ≠BEGIN≠ K.=1.,≠GO TO≠ M2., ≠END≠.,
    ≠IF≠ DB(/DFB(/I/),2/) ≠NOT EQUAL≠ 0 ≠THEN≠
    ≠BEGIN≠
      ≠IF≠ K ≠NOT EQUAL≠ DB(/DFB(/I/),2/) ≠THEN≠
      ≠BEGIN≠
        K.=DB(/DFB(/I/),2/).,≠GO TO≠ M1.,
      ≠END≠.,
    ≠END≠.,
  ≠END≠.,
AUSI.,
≠COMMENT≠ NORMIERUNG DER NEUEN FAERBUNG.,
⋮
≠END≠ UMFARB.,
```

The variables in the program signify the following:

N   : size of the boundary coloring (*number*)
B   : boundary coloring (*vector*)
ANZ   : number of points of contact of the Kempe sectors (*even number*)
KOMPL : color exchange (*4-tuple* $(w, ?, ?, ?)$)
KET   : list of points of contact (ANZ+2-*tuple*)
U   : ANZ/2 − 1

DB : block decomposition ($U \times 2$-*matrix*)

DFB : list of blocks in which the colors will be interchanged (U-*tuple*)

NORM : *will be needed for the transition (which is no longer represented) to the corresponding essential boundary coloring.*

The ANZ+2-tuple KET is explicitly defined for a boundary coloring of size N as follows:

KET(/0/)   = 0

KET(/K/)   = K-th point of contact, for $K \in \{1, \ldots, ANZ\}$,

$$\text{KET}(/\text{ANZ}+1/) \; = \; \begin{cases} 0, & \text{in the case that there are N points of contact,} \\ N, & \text{otherwise.} \end{cases}$$

The significance of the number U is as follows: A block decomposition, of course, contains $U + 2$ blocks but is already determined by U blocks. In addition, only block selections with at most U blocks are of interest for the upcoming Kempe interchanges that are relevant. In order to describe the matrices DB and the tuple DFB, another *ordering* or *numbering* of the blocks will be required. For that, we start with an essential boundary coloring a and a color-pair choice w. We first number the Kempe sectors from 1 to q according to the sequence of their smallest elements. In doing so, the w-sectors will have odd numbers, and the $\overline{w}$-sectors will have even numbers. A block B is determined by the number of the Kempe sectors whose union forms the block in question. We denote by $|B|$ the number of Kempe sectors in B and by $\min B$ the smallest number assigned to any of its sectors. If B' and B'' are blocks of a block decomposition, then we set $B' < B''$ if one of the following two conditions holds:

1. $|B'| < |B''|$,

2. $|B'| = |B''|$ and $\min B' < \min B''$.

In this way, a linear ordering is established on the set of blocks of a block decomposition. The blocks will be numbered in accordance with this ordering, beginning with the number 1. Table 6.2 on page 201 contains examples. This numbering deviates somewhat from Dürre's original numbering, but it fulfills the purpose of making the part of the program shown above more understandable.

In order to define the size of the matrices DB, it suffices to note that with boundary colorings of size 6, block decompositions of U blocks are determined by at most two Kempe sectors. Thus the rows of a matrix DB represent the first U blocks of a block decomposition. In a row, the first entry contains the number of the first Kempe sector belonging to the block. The second entry contains the number of the second Kempe sector belonging to the block or the number 0 (in the case where the block contains only one Kempe sector).

Finally, only the tuple DFB remains to be defined. It contains the numbers of the blocks that will be involved in the Kempe interchange and that will be filled with zeros of length U.

6.5 A-, B-, and C-Reducibility

Out of the preliminary discussions on reducibility comes the aware-
ness that one more special notion will be needed—one that cannot
be encompassed within a short definition. The treatment of the
Birkhoff diamond in Section 6.1 can serve as an introductory example in which we have already mentioned the idea of a "reducer"
(see page 181). More generally, we define:

Definition 6.5.1
Let C be a configuration with the outer vertices $\mathbf{x}_1, \ldots, \mathbf{x}_r$ given in
cyclic order. A pair (S, σ) consisting of a graph S and a surjective
mapping σ from the set of outer vertices of C to the set of outer
vertices of S is said to be a *reducer* for C if S has fewer vertices than
C and the following conditions hold:

1. σ preserves the property of being a neighbor. This means that for
   all $j \in \{1, \ldots, r\}$, $\sigma(\mathbf{x}_j)$ and $\sigma(\mathbf{x}_{j+1})$ are neighbors and in particular
   are distinct neighbors (with respect to addition modulo r).
2. The original distinct outer vertices of S with respect to σ cannot
   mutually overlap. In other words, for $j_1, j_2, k_1, k_2 \in \{1, \ldots, r\}$,
   where

$$\sigma(\mathbf{x}_{j_1}) = \sigma(\mathbf{x}_{j_2}) \neq \sigma(\mathbf{x}_{k_1}) = \sigma(\mathbf{x}_{k_2}),$$

then the following ordering is not possible:

$$j_1 < k_1 < j_2 < k_2.$$

We observe that a graph S that is the component of a reducer can
even contain bridges and final edges but is in any case connected.

So why talk about reducers at all? We consider a minimal triangulation G that contains a configuration C for which a reducer (S, σ) is
given. Then we construct a graph G_σ that has fewer vertices than G
in the following way. First of all, as was done in the discussion about
chromodendra, we remove from G the inner vertices of C and the
edges incident with them, thereby obtaining a graph G'. This new
graph has an exceptional face (in this case, a bounded face L) that is
bordered by an r-gon. By the Schoenflies theorem (Theorem 2.2.7),
we can assume that L is bounded by a regular r-gon. In the next
step, we join every pair of vertices that are mapped by σ to the same

value with a diagonal. Since the original pairs of these vertices do not overlap with respect to σ, these pairs of diagonals do not intersect each other. However, they may abut at a common end point. Thus we are able to add these diagonals as edges to our graph. Of course, it can happen that multiple edges are created by this process. This is precisely the case if two outer vertices of C that have the same image under σ are joined by an edge in G'. This cannot be an outer edge of C. In order to preclude this phenomenon from happening in general, we have introduced the notion of a "proper" embedding (Definition 5.4.9). We say that C is σ-*properly* embedded in G if two outer vertices of C that have the same image under σ are not neighbors in G. We now assume that this property holds and denote by G'' the graph that is produced by the addition of the aforementioned diagonals. However, simultaneously, we geometrically contract each of these new edges to one of its points (see page 148/9). In this way, they disappear. Denote by G''' the graph resulting from this process. The now deformed boundary of the exceptional face of G' is then mapped homeomorphically onto the boundary of the unbounded face of the graph S. The inverse mapping can be extended to an embedding of all of S into the plane. This means that:

1. The images of the inner vertices of S lie in faces of G'''.

2. If the images of the end points of an edge B of S are joined by an edge B' in G''', then B will be mapped to B'.

3. The images of the remaining edges of S, possibly with the exception of one or both end points, lie entirely in faces of G'''.

These conditions guarantee that we are able to add the images of the vertices and edges of S to G''', unless, of course, they are already vertices or edges of G'''. All of this can be done without altering the properties of the graph. We obtain a graph G_σ in which S is embedded in such a way that it replaces the configuration C in the original minimal triangulation G. We extend the mapping σ to a mapping σ' of the vertex set of G' into the vertex set of G_σ by which each vertex of G' that is not an outer vertex of C is mapped to the corresponding vertex in G_σ. On the basis of the definition of a reducer, the graph G_σ has fewer vertices than G. It therefore has an admissible vertex 4-coloring χ. The composition $\chi \circ \sigma'$ is then an admissible vertex 4-coloring of G'. The reducibility of C would, in any case, be assured only if C could be properly embedded in a minimal triangulation and if each such resulting vertex 4-coloring could be extended to the inner vertices of C. At this point in the discussion, we are already very close to the notion of A-reducibility. Since in terms of the aforementioned extendibility it depends only upon the coloring of the outer vertices of C, it suffices to consider the composition $\chi \circ \sigma$ that is determined by the values of χ on the outer vertices of S. The coloring of the outer vertices of S cannot, however, be arbitrarily chosen. They must be admissible and be extendible, if necessary, to the inner vertices. Thus we can make the transition to the set $\Psi(S)$ of admissible vertex 4-colorings of S. We say that a boundary coloring of C is *σ-compatible* if interpreted as a mapping it is of the form $\chi \circ \sigma$, where $\chi \in \Psi(S)$. The set of σ-compatible boundary colorings of C will be denoted by $\Phi(r, \sigma)$. In general, $\Phi(r, \sigma)$ is a proper subset of $\Phi(r)$. In our example of a reducer for the Birkhoff diamond, only 6 of the 31, in total, boundary colorings of size 6 are σ-compatible (see page 182).

Definition 6.5.2
A configuration C is said to be *A-reducible* if it has a reducer (S, σ) satisfying the following conditions:
1. C can be σ-properly embedded only in a minimal triangulation.
2. Each σ-compatible boundary coloring is directly color-extendible. In other words,

$$\Phi(r, \sigma) \subset \Phi_0(C).$$

The simplest example of A-reduction is the graph-theoretical version of our first proof of the reducibility of the 4-star (Theorem 4.5.4). In addition, most of the earlier-mentioned results of Birkhoff can be achieved using A-reduction. In any case, no reducer is known by which the Birkhoff diamond is A-reducible. We now give another class of examples, that were discovered by Franklin [FRANKLIN 1922].

Theorem 6.5.3
A configuration whose core consists of one n-vertex together with $n - 1$ neighboring 5-vertices is A-reducible. ∎

Proof Let C be a configuration with the given properties. Denote by y_0 the unique inner vertex of degree n. It has one leg, whose other end point will be denoted by x_0. The remaining neighbors of y_0 are the 5-vertices $y_1, \ldots, y_{n-1}$ labeled cyclically in such a way that y_1 and y_{n-1} are also neighbors of x_0. The other outer vertices will be denoted by $x_1, \ldots, x_n$ so that the following conditions hold: x_1 is a neighbor of x_0 and y_1; x_j is a neighbor of y_{j-1} and y_j for $j \in \{2, \ldots, n-1\}$; and, finally, there exists a vertex x_n that is a neighbor of y_{n-1} and x_0. Thus C has ring size $n + 1$. Moreover, as the secondary neighbors in a minimal triangulation also form a ring (Theorem 6.1.9), $(x_0, \ldots, x_n)$ is always a simple closed chain. In other words, C is always properly embedded in a minimal triangulation. Now it is necessary to distinguish between the various cases:

First, let n be even. We obtain a reducer (S, σ) in which we deform C by contracting the inner vertices and the outer vertices indexed with odd numbers to a vertex z. Since the vertices z, x_0, and x_n form a triangle in S, they must be differently colored in each admissible vertex 4-coloring of S. We can assume that they take on the colors 0, 1, 2 in that order. Thus the following structure emerges for the σ-compatible boundary colorings of C:

Vertex	x_0	x_1	x_2	$\ldots$	x_{2k-1}	x_{2k}	$\ldots$	x_{n-1}	x_n
Color	1	0	f_2	$\ldots$	0	f_{2k}	$\ldots$	0	2

where $f_{2k} \in \{1, 2, 3\}$ for all $k \in \{1, \ldots, \frac{n}{2} - 1\}$. Each such coloring can be extended to the inner vertices of C in the following way:

- y_0 has the color 0.

- y_{n-1} has x_0, y_0, y_{n-2}, x_{n-1}, and x_n as neighbors. Therefore, the colors 0, 1, 2 are already used. Hence, only the color 3 remains for y_{n-1}.

- If $f_{2k} \neq 1$ for all relevant k, then the remaining inner vertices with an even-number index take on the color 1. The vertices y_{2k-1} take on the colors 2 or 3 depending on whether $f_{2k} = 3$ or 2, respectively.

- Otherwise, there exists a smallest index k_0 with $f_{2k_0} = 1$.

- Now the vertices y_{2k} for $k \in \{1, \ldots, k_0 - 1\}$ take on color 1, and the vertices y_{2k-1} take on color 2 or color 3 depending on whether $f_{2k} = 3$ or 2, respectively. This step is omitted if $k_0 = 1$.

- Then we color y_{n-2}, y_{n-3}, y_{2k_0} in order of sequence. Each such y_j has y_0, y_{j+1}, x_{j+1}, x_j, and y_{j-1} as neighbors. When y_j appears in the series, four neighbors have already been colored. However, since either j or $j+1$ is odd, two of these neighbors have, in any case, the same color 0. Therefore, at most 3 colors are needed for the neighbors. One color remains unused, and y_j can be colored with this remaining color.

- The vertex y_{2k_0-1} still remains to be colored. All of its neighbors are already colored. Two of them, y_0 and x_{2k_0-1}, are, of course, colored with the color 0. Two of them, y_{2k_0-2} and x_{2k_0}, are colored with color 1. The fifth neighbor takes on a third color. There is now exactly one color free with which y_{2k_0-1} can be colored.

For odd n, we obtain a suitable reducer (S, σ) for C by the following contractions:

- The vertices x_0, y_1, and x_2 will be collapsed to a vertex z_1.

- The outer vertices having odd index > 1 and the remaining inner vertices will be collapsed to a vertex z_0.

It suffices to consider admissible vertex 4-colorings of S by which z_0 is colored with the color 0, x_1 with 0 or 1, and z_1 with 2. Now the following structure emerges for the coloring of the outer vertices of C.

Vertex	x_0	x_1	x_2	x_3	x_4	...	x_{2k-1}	x_{2k}	...	x_{n-1}	x_n
Color	2	0, 1	2	0	f_4	...	0	f_{2k}	...	f_{n-1}	0

where $f_{2k} \in \{1, 2, 3\}$ for all $k \in \{2, \ldots, \frac{n-1}{2}\}$. This coloring as well can be extended to the inner vertices of C. The construction proceeds in a manner similar to the first case.

- First of all, we color y_0 again with the color 0. Then, however, y_1 is colored with color 3 and y_2 with color 1.

- If $f_{2k} \neq 1$ for all relevant k, the remaining inner vertices with even number indices also take on the color 1. The vertices y_{2k-1} take on the color 2 or 3 depending on whether $f_{2k} = 3$ or 2, respectively.[6]

---

[6]The reader is left to provide a proof for this and proofs for the following parts by referring back to the first case.

- Otherwise, there exists a smallest index k_0 with $f_{2k_0} = 1$.

- Now the vertices $\mathbf{y}_{2k}$ for $k \in \{2, \ldots, k_0 - 1\}$ obtain the color 1, and the vertices $\mathbf{y}_{2k-1}$ take on the color 2 or 3 depending on whether $f_{2k} = 3$ or 2, respectively. This step is omitted if $k_0 = 2$.

- Then we color the vertices $\mathbf{y}_{n-1}$, $\mathbf{y}_{n-2}$, $\mathbf{y}_{2k_0}$ in order of sequence. Each such $\mathbf{y}_j$ has $\mathbf{y}_0$, $\mathbf{y}_{j+1}$ (which is $\mathbf{x}_0$ in the case $j = n - 1$), $\mathbf{x}_{j+1}$, $\mathbf{x}_j$, and $\mathbf{y}_{j-1}$ as neighbors. When $\mathbf{y}_j$ appears in the series, four neighbors have already been colored. However, since either j or $j+1$ is odd, two of these neighbors have in any case the same color 0. Therefore, at most 3 colors are needed for the neighbors. One color remains free. With this, $\mathbf{y}_j$ can be colored.

- The vertex $\mathbf{y}_{2k_0-1}$ remains to be colored. All of its neighbors are already colored. Two of them, $\mathbf{y}_0$ and $\mathbf{x}_{2k_0-1}$, are colored with color 0. Two of them, $\mathbf{y}_{2k_0-2}$ and $\mathbf{x}_{2k_0}$, are colored with color 1. The fifth neighbor takes on a third color. One color remains free. That color can be used for $\mathbf{y}_{2k_0-1}$. □

So much for A-reduction. The descriptions of B- and C-reduction are now very easy.

Definition 6.5.4
Let C be a configuration, and let (S, σ) be a reducer for C in such a way that C can be σ-properly embedded only in a minimal triangulation. The configuration C is said to be:

- *B-reducible* if each σ-compatible boundary coloring is either good from the outset or good from stage 1. This means that

$$\Phi(r, \sigma) \subset \Phi_1(C).$$

- *C-reducible* if each σ-compatible boundary coloring is good from some stage or other. In other words,

$$\Phi(r, \sigma) \subset \overline{\overline{\Phi}}(C).$$

An example of B-reduction comes from the Birkhoff diamond. We have already observed (page 210) that the reducer used by Birkhoff (and by us) produces six σ-compatible boundary colorings. The first five (in the order given in the table on page 182) are good from the outset, and the last is good from stage 1.

As an example of C-reduction, we take a configuration that was first studied by Chojnacki–Hanani [CHOJNACKI 1942].

Theorem 6.5.5

The configuration C whose core consists of one 8-vertex and five of its neighbors, one following the other and each with degree 5, is C-reducible. ■

A Sketch of the Proof We have one inner 8-vertex $\mathbf{y}_0$ and a chain $(\mathbf{y}_1, \ldots, \mathbf{y}_5)$ of neighbors of $\mathbf{y}_0$ where each is an inner vertex and each is of degree 5. The vertex $\mathbf{y}_0$ has three legs, whose other end points are denoted by $\mathbf{x}_0$, $\mathbf{x}_1$, $\mathbf{x}_2$ in such a way that $\mathbf{x}_2$ is a neighbor of $\mathbf{y}_1$ and $\mathbf{x}_0$ is a neighbor of $\mathbf{y}_5$. There exist another six outer vertices, which we will denote by $\mathbf{x}_3, \ldots, \mathbf{x}_8$ so that the 9-tuple $(\mathbf{x}_0, \ldots, \mathbf{x}_8)$ forms a closed chain. As in the configurations discussed above, which were studied by Franklin, everything concerning the embedding in a minimal triangulation happens within the secondary neighborhood of a vertex. Therefore, this configuration C also always occurs only in a properly embedded form (Theorem 6.1.9).

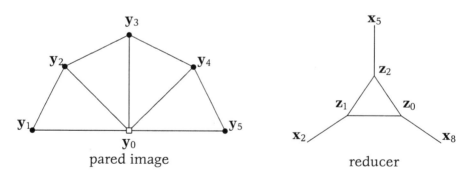

pared image   reducer

Chojnacki–Hanani found a reducer (S, σ) for C using the following contractions:

- The vertices $\mathbf{x}_0$, $\mathbf{y}_5$, $\mathbf{y}_4$, and $\mathbf{x}_7$ are collapsed to a vertex $\mathbf{z}_0$.
- The vertices $\mathbf{x}_1$, $\mathbf{y}_0$, $\mathbf{y}_1$, and $\mathbf{x}_3$ are collapsed to a vertex $\mathbf{z}_1$.
- The vertices $\mathbf{x}_4$, $\mathbf{y}_2$, $\mathbf{y}_3$, and $\mathbf{x}_6$ are collapsed to a vertex $\mathbf{z}_2$.

Since the vertices $\mathbf{z}_0$, $\mathbf{z}_1$, and $\mathbf{z}_2$ form a triangle in S, they must be differently colored in each admissible vertex 4-coloring of S. We can assume that they take on the colors 0, 1, 2 in order of sequence. Thus the following structure for the σ-compatible boundary colorings of C emerges.

Vertex	x_0	x_1	x_2	x_3	x_4	x_5	x_6	x_7	x_8
Color	0	1	f_2	1	2	f_5	2	0	f_8

where $f_2 \in \{0, 2, 3\}$, $f_5 \in \{0, 1, 3\}$, $f_8 \in \{1, 2, 3\}$. We obtain 27 σ-compatible boundary colorings, of which 18 prove to be directly color-extendible. The remainder are good from some higher stage onwards. □

Remark: At a later date, it was established that the configuration discussed above is even D-reducible [APPEL, HAKEN, and KOCH 1977, Figure 7-1]. ◇

How then are the various notions of reducibility interconnected? Obviously, A-reduction is a special case of B-reduction, and B-reduction is a special case of C-reduction. However, D-reduction can also be thought of as a special kind of C-reduction. Suppose we are given a configuration C. Choose the graph S to be the bounding circuit of C. Let the mapping σ be the identity mapping. Since this particular σ identifies no outer vertices of the configuration, C always occurs only as being σ-properly embedded in a minimal triangulation. Hence,

$$\Phi(r, \sigma) = \Phi(r).$$

In this way, the notion of C-reduction is the most encompassing of all the reduction concepts.

For each configuration C of ring size r, there exist only finitely many reducers (S, σ). As the number of vertices of a reducer is bounded from above by the number of vertices of the configuration and since there exists only a finite number of graphs for each fixed number of vertices, there can exist only finitely many possible choices for the graph S. However, if C and S are fixed, then there also exists only a finite number of possibilities for the mapping σ. This means that the entire set $\Phi(r, \sigma)$ can be realized algorithmically. These sections on A-, B-, and C-reducibility are therefore in principle verifiable using a computer. What remains is the problem of σ-proper embeddings. It is only natural that a significant portion of the original proof has been devoted to this problem [APPEL, HAKEN, and KOCH 1977, Section 3].

To close this chapter, we make a few more remarks about inner vertices of a reducer—more precisely, of the graph S that occurs as the first component of a reducer. Heesch, in his book [HEESCH 1969], wrote that, "in practice," only reducers without inner vertices actually crop up. However, Koch, in his dissertation [KOCH 1976], proved that reducers having inner vertices would be required in the reduction of a C-reducible configuration of ring size 10 and in the reduction of three C-reducible configurations of ring size 11. These configurations actually occur in the proof of the Four-Color Theorem. Using the numbering given by [APPEL, HAKEN, and KOCH 1977], they occur in the following diagrams: Figures 3-10, 3-28 (page 507), 7-27 (page 511) and 17-9 (page 521). In the appendix to [APPEL, HAKEN, and KOCH 1977], published in 1989, a reducer with three inner vertices for Figure 16-14 with ring size 12 [APPEL and HAKEN 1989, page 483] was also mentioned. This was discovered by Allaire. No hint was given, however, whether it could have been done in another way. Appel, Haken, and Koch noted that the C-reducibility of Figure 16-14 produced as a consequence a significant simplification of their list of unavoidable configurations.

Here we show the C-reducible configuration 3-28 having ring size 10 together with a reducer (communicated to us by Koch) that has two inner vertices.

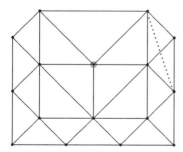 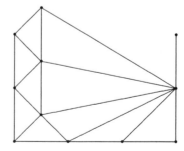

The dotted line segment in the unpared image of configuration 3-28 (on the left) is collapsed to a single vertex in the construction of the reducer (on the right).

A reducer for this configuration was also mentioned by Allaire and Swart [ALLAIRE and SWART 1978]. It consists in the removal in the original configuration of the vertex that is especially marked by an additional circle in the sketch above and all edges incident with it. Therefore, the reducer has five inner vertices.

In the new proof of the Four-Color Theorem [ROBERTSON et al. 1997], Robertson, Sanders, Seymour, and Thomas are able to restrict themselves to very specific kinds of reducibility. They need only determine *D*-reducibility and *C*-reducibility where reducers are obtained from the configurations in question by the contraction (see Definition 5.2.8) of a *sparse set* of at most four edges. A sparse set of edges is defined as follows:

Definition 6.5.6

A set $\mathcal{B}$ of edges of a configuration $\mathcal{C}$ is called *sparse* if it consists only of inner edges of $\mathcal{C}$ no two of which belong to the boundary of the same triangle.

To prove reducibility of their 633 configurations, they found 3-colorings of edges easier to manipulate than 4-colorings of vertices. Therefore, in view of the famous equivalence result of Tait (Theorem 4.8.2), for their computer programs they chose instead to examine edge 3-colorings.

7 The Quest for Unavoidable Sets

7.1 Obstructions and a "Rule of Thumb"

It is Heinrich Heesch who must be credited with compiling a number of facts about how to search for unavoidable sets of reducible configurations. He provided clues as to which configurations one could most likely ignore. Initially, it was a matter of three "obstructions."[1] Even up to now, none of the known methods of reduction have been successful in reducing a configuration whose inner vertices all have at least degree 5 if it "essentially" contains one of the following internal structures:

1. An inner vertex with more than three legs.

2. An articulation with more than two legs. (This explains an earlier observation on page 166 that configurations with articulations

---

[1] Here we use quotation marks instead of italics to point out that it has more to do with a "physical" phenomenon than with precise mathematical definitions and theorems.

having more than two legs can remain outside of the realm of consideration.)

3. A *dangling 5-couple*, which is a figure consisting of two neighboring inner 5-vertices both of which adjoin the same third distinct inner vertex. The vertices belonging to a dangling 5-couple have exactly three legs each.

An obstruction "essentially" arises in a configuration if the configuration is not obtainable by extending a reducible configuration, out of which this obstruction and eventually others may be realized. That leads to the following simple test for reducibility.

Suppose we have a configuration C having no inner vertices of degree less than 5. One makes the transition, in so far as is possible, to other configurations by altering the pared images in the following manner:

1. One discards all vertices with more than three legs and all edges incident with them.

2. One discards all articulations with at least three legs and those edges incident with them.

3. One eliminates all dangling 5-couples and all edges incident with at least one vertex of such a couple.

At each stage, one obtains either the "empty" graph, which is the only graph whose sets of vertices and edges are both empty, or the pared images of one or *several* configurations. The word "several" is, in this instance, used ambiguously. On the one hand, by removing an articulation, the pared image in question decomposes into several components. On the other hand, the configuration, in the case of an articulation having more than two legs, is not uniquely determined by its pared image (see the remark on page 166).

This procedure terminates after finitely many steps. If in the final analysis one obtains either the empty graph or the pared images of a single or several configurations all known to be irreducible, then the original configuration C is, to a high degree of probability, also irreducible (at any rate, up to now no counterexample has been found).[2] On the other hand, if one ends up with pared images of a number of configurations so that reducibility can be proved for at least one of them, then no conclusion as to the reducibility or the irreducibility of C is possible.

It may be helpful to clarify these observations with a few examples, the simplest of which are the following:

1. The core of the Birkhoff diamond (see page 155) has two vertices each with two legs and two other vertices each with three legs. In the case of the latter two vertices, neither is an articulation, nor are the vertices adjacent. One encounters absolutely no obstructions. Therefore, the test ends where it begins—with a reducible configuration.

2. All inner vertices of both Wernicke configurations (see page 167) have at least four legs. In both cases, the test, after the first step, results in the empty graph and with a very high degree of probability indicates irreducibility.

3. The A-reducible Franklin configurations with n inner vertices (Theorem 6.5.3) also have neither articulations nor dangling 5-couples. There does exist one inner vertex with one leg and two others, each having three legs. All of the other inner vertices are 2-legged vertices.

4. The C-reducible (even D-reducible) Chojnacki configuration (see Theorem 6.5.5 and the following remark), as well, contains neither articulations nor dangling 5-couples. Three of its inner vertices have two legs, and the other three have three legs each.

To his obstruction theory, which was designed to detect irreducibility, Heesch added the following "rule of thumb." It enables one to ascertain with a greater degree of likelihood whether a configuration is reducible.

---

[2]The 4-star does not satisfy the hypothesis that all inner vertices must have degree at least 5. For this reason, it does not serve as a counterexample of a reducible configuration with one obstruction: namely, one inner vertex with 4 legs.

A configuration having ring size n with m inner vertices but no obstructions is more likely to be reducible when

$$m > \frac{3n}{2} - 6. \qquad (7.1.1)$$

A complete justification for this is found in Aigner's book [AIGNER 1984]. We state only a few examples:

- In the case of the 4-star, $m = 1$ and $n = 4$, and so $\frac{3n}{2} - 6 = 0 < m$.

- For the k-star where $k \geq 5$, $m = 1$, $n = k$. In this case, $\frac{3n}{2} - 6 \geq \frac{3}{2} > m$.

- With the Birkhoff diamond, $m = 4$ and $n = 6$. Therefore, $\frac{3n}{2} - 6 = 3 < m$.

Another kind of obstacle is the size of the ring. In this case, it is not a question of the reducibility of a configuration but of the ability to verify (or not) C-reducibility in a "tolerable" number of steps. This is a matter of having at one's disposal a computerized environment that has hardware with adequate memory capacity and the availability of enough computing time. One must imagine that simply by raising the size of the ring by 1, not only the computing time but also the memory requirements almost quadruple. This is alarming if one bears in mind that Dürre's program for a special configuration of ring size 14 took no less than 26 hours. Of course, that involved a particularly nasty configuration. Appel and Haken averaged only about 25 minutes computing time for configurations of ring size 14 [APPEL and HAKEN 1989, page 8]. However, this value skyrockets to a computing time requirement of about 100 hours for ring sizes 18. For this reason, one is obliged to search for unavoidable sets of configurations having restricted ring sizes.

On the other hand, increasing the ring size also increases the probability of reducibility. This one can deduce from Heesch's "rule of thumb." If one takes the smallest "neighborhood" of a configuration with ring size n and m inner vertices (meaning that one considers a configuration whose core is precisely the given configuration), then one can roughly estimate that the corresponding ring size grows linearly but the number of inner vertices increases quadratically.[3] Thus one gets closer and closer to condition (7.1.1) of Heesch's "rule of thumb."

Between these two considerations, limitations of computing time and memory on the one hand and Heesch's "rule of thumb" on the other, one must find an optimal balance. The first attempts by Appel and Haken were limited to 2000 configurations with possible ring sizes of at most 16. In fact, their unavoidable set of 1834 reducible configurations contained only elements having ring size at most 14. The number of configurations of the various ring sizes are listed below:

Ring Size	≤ 8	9	10	11	12	13	14	≤ 14
Number	7	8	35	89	334	701	660	1834

7.2 Discharging Procedures

In the continued search for suitable unavoidable sets of configurations, Heesch devised a clever technique [HEESCH 1969] that Haken called a "discharging procedure" [HAKEN 1973]. Haken's suggestive terminology, which has ultimately been adopted, is reminiscent of the physical notion of the discharge of positively charged ions, called cations. Heesch saw a connection with the total positive curvature of the sphere and called the magnitudes that arose out of this process "curvatures." Haken then named them "charges." Mayer independently from Heesch, but somewhat later, developed similar ideas. He chose to use the word "compensation" for the notion of a charge [MAYER 1975].

The discharging procedures consist of a tricky interpretation of inequality (4.6.6) that one, in the cases still to be examined (that is, the normal graphs having no vertices with degree smaller than 5),

---

[3]The circumference of a circle, $2\pi r$, is dependent linearly on its radius r, which means that r is to the power one in the formula for circumference. However, r is to the power 2 in the formula for the area of a circle, πr^2.

can write as an equation in the following way:

$$v_5 - v_7 - 2v_8 - \cdots - (s-6)v_s = 12, \qquad (7.2.1)$$

where s denotes the maximum degree of the vertices involved. To begin with, each vertex is given the initial charge of $60 \cdot (6-$ degree of vertex). Heesch chose the factor 60 in order to avoid fractions as much as possible. From identity (7.2.1), it follows that the total charge of the system is $+720$, in which, however, only the degree 5 vertices are positively charged. During the process of a "discharging procedure," the charges are removed in such a way that the degree 5 vertices give off positive charges, thereby becoming discharged. The total charge, however, does not change. In addition, after the discharging process, some positively charged vertices must remain. It is from these positively charged vertices that one can infer the existence of an unavoidable set of configurations.

We explicitly point out once again that in this chapter we are limiting our study to normal graphs with no vertices having degree less than 5. This means that we will not be including the 3-star and the 4-star in our list of unavoidable sets. The following linguistic convention has been adopted.

- Vertices of degree 5 or 6, that is, those vertices with nonnegative charge, are said to be *minor* vertices. Vertices of degree greater than 6 are called *major* vertices.

The very simplest example of a discharging procedure corroborates Wernicke's result that the 5-5-chain and the 5-6-chain (see page 167) form an unavoidable set. The proof goes as follows. Every 5-vertex releases a discharge of 12 units to every neighboring major vertex. One can then derive that 5-vertices with only major vertices as neighbors are completely discharged. Vertices of degree $d \geq 8$ have absorbed at most a charge of $12 \cdot d$. Out of the initial charge of $60 \cdot (6 - d)$ one calculates that the charge is now less than or equal to

$$60(6-d) + 12d = 360 - 48d \leq 360 - 384 = -24$$

and is thereby negative. Since the total charge is positive, there must therefore be either at least one 7-vertex that is positively charged or a 5-vertex that is not completely discharged. In the latter case, one has either two neighboring 5-vertices or a 6-vertex adjoined to a 5-vertex. The first case can occur only if a 7-vertex has at least six neighboring 5-vertices, and of these at least two of them must also be adjacent to one another.

This example of a discharging procedure is, unfortunately, not very helpful. The resulting configurations defy attempts at reduction because they contain obstructions (see page 221, example 2). At this juncture, we would like to present yet another interesting example that was given by Haken. With this example, he verified a result of Chojnacki–Hanani that we have already cited (see page 185).

Theorem 7.2.1

The 6-star, the 7-star, the Birkhoff diamond, the Chojnacki configuration (Theorem 6.5.5), and the Franklin configurations with 9, 10, *and* 11 *inner vertices (Theorem 6.5.3) together form an unavoidable set.* ∎

Proof It will be shown that by using a minimal triangulation that contains *none* of the configurations listed above a complete discharge is possible. This means a discharge after whose implementation no vertices of positive charge remain. This is in contradiction to the fact that throughout the entire discharging process the total charge remains the same.

Let **G** be such a minimal triangulation. Major vertices can have neighbors of degree 5. These we subdivide into two types:

- The neighboring vertex **x** of the major vertex **y** is a 5-*neighbor of type 1* if **x** is a 5-vertex and if in the ring of neighbors of **y** there exists exactly one 5-vertex that is a neighbor to **x**.

- The neighboring vertex **x** of the major vertex **y** is a 5-*neighbor of type 2* if **x** is a 5-vertex but is not of type 1. This means that in the ring of neighbors of **y**, the number of existing neighbors of **x** with degree 5 totals either 0 or 2.

We denote by E_5 the set of 5-vertices and by E_g the set of major vertices of **G**. Then we define three functions $p_1 : E_g \to \mathbb{N}_0$, $p_2 : E_g \to \mathbb{N}_0$, and $w : E_5 \times E_g \to \mathbb{N}_0$ as follows:

For all major vertices **y**,

- $p_1 = p_1(\mathbf{y}) =$ the number of 5-neighbors of **y** of type 1,
- $p_2 = p_2(\mathbf{y}) =$ the number of 5-neighbors of **y** of type 2.

For all 5-vertices $\mathbf{x}$ and all major vertices $\mathbf{y}$,

$$w = w(\mathbf{x}, \mathbf{y}) = \begin{cases} 60 & \text{if } \mathbf{x} \text{ is a 5-neighbor of } \mathbf{y} \text{ of type 1,} \\ 120 & \text{if } \mathbf{x} \text{ is a 5-neighbor of } \mathbf{y} \text{ of type 2,} \\ 0 & \text{otherwise.} \end{cases}$$

In contrast to the previous example, here the discharging takes place from the major vertices to the 5-vertices. This means that there will be a shifting of negative charges. Each 5-vertex $\mathbf{x}$ receives from each major vertex $\mathbf{y}$ of degree d the negative charge

$$l = l(\mathbf{x}, \mathbf{y}) = w \cdot \frac{6 - d}{p_1 + 2p_2},$$

in which we interpret the right-hand side as being 0 if $\mathbf{y}$ has no 5-neighbors at all. After this process, the resulting charges are as follows:

- For 5-vertices $\mathbf{x}$: $q(\mathbf{x}) = 60 + \sum l(\mathbf{x}, \mathbf{y})$, where the summation runs over all major vertices $\mathbf{y}$.
- For d-vertices $\mathbf{y}$, $d \geq 8$: $q(\mathbf{y}) = 60(6 - d) - \sum l(\mathbf{x}, \mathbf{y})$, where the summation runs over all 5-vertices $\mathbf{x}$.

That the number 8 is the lower bound for d follows from the assumption that there are no 6-vertices and no 7-vertices. The final charge of the major vertex $\mathbf{y}$ of degree d can be immediately and explicitly stated:

$$q(\mathbf{y}) = \begin{cases} 60 \cdot (6 - d) & \text{if } \mathbf{y} \text{ has no 5-neighbor,} \\ 0 & \text{otherwise.} \end{cases}$$

Consequently, it is never positive.

It remains to be shown that $q(\mathbf{x}) \leq 0$ for all 5-vertices $\mathbf{x}$. We first prove the following inequality:

If $\mathbf{x}$ is a 5-vertex that is a neighbor of a major vertex $\mathbf{y}$ of degree d, then

$$|l(\mathbf{x}, \mathbf{y})| \geq \frac{w(\mathbf{x}, \mathbf{y})}{4}, \tag{7.2.2}$$

or

$$4 \cdot (d - 6) \geq p_1 + 2p_2.$$

Thus we have to distinguish between several distinct cases:

$d = 8$: Since the Chojnacki configuration is, by assumption, excluded, the only possibilities for p_1 and p_2 are the following:

p_1	0	2	4
p_2	≤ 4	≤ 3	2

Hence, $p_1 + 2p_2 \leq 8$ always holds, which was what we had to show.

$d = 9, 10, 11$: Since the corresponding Franklin configurations (again by assumption) are excluded from consideration we have in any case that $p_1 + p_2 \leq d - 2$ and therefore that $p_2 \leq d - 4$. Hence,

$$p_1 + 2p_2 = (p_1 + p_2) + p_2 \leq 2d - 6 < 4 \cdot (d - 6).$$

$d \geq 12$: From $p_1 + p_2 \leq d$ and $2d \geq 24$ it follows that

$$p_1 + 2p_2 \leq 2d \leq 4d - 24 = 4 \cdot (d - 6).$$

Now we point out that a 5-vertex has at least two major vertices as neighbors. In the case of having precisely two, these vertices cannot themselves be adjacent. Otherwise, the Birkhoff diamond would be in G, which contradicts the initial hypothesis. From this, it follows that for a 5-vertex $\mathbf{x}$ there are the following three possibilities:

- $\mathbf{x}$ has exactly two major neighboring vertices. Then $\mathbf{x}$ is a 5-neighbor of type 2 of both of them and inherits negative charges from both, each charge having at least a value of 30 (let $w = 120$ in the inequality (7.2.2)). Hence, the resulting charge $q(\mathbf{x}) \leq 0$.

- $\mathbf{x}$ has exactly three major neighboring vertices and is a 5-neighbor of type 2 of exactly one of them. Therefore, because of inequality (7.2.2), $\mathbf{x}$ receives the following negative charges: one charge of at least 30 and two of at least 15. Again, it follows that $q(\mathbf{x}) \leq 0$.

- $\mathbf{x}$ has at least four major neighboring vertices. In this case, it receives from each a negative charge of at least 15 because of inequality (7.2.2) and the fact that $w \geq 60$. This too yields $q(\mathbf{x}) \leq 0$. $\qquad \square$

Corollary 7.2.2 (Chojnacki–Hanani/Heesch Theorem)
A minimal triangulation contains at least one 6-vertex or one 7-vertex. ■

Proof   In the previous theorem, all of the other configurations listed in the unavoidable set are reducible.   □

The outstanding and independent contribution of Appel and Haken to the proof of the Four-Color Theorem consisted in the development of efficient discharging procedures (see page 34). At first, they endeavored to construct unavoidable sets of configurations by avoiding Heesch's obstructions (see page 219), thereby deferring the question of true reducibility. Before embarking on the joint work with Appel, Haken had found an unavoidable set of 68 configurations in which at most one inner vertex had more than three legs. To achieve a more exact estimate of the orders of magnitude of possible unavoidable sets of reducible configurations, Appel and Haken, together, with the aid of a computer, looked for an unavoidable set consisting of configurations that were called "geographically good."

Definition 7.2.3

A configuration is called *geographically good* if no vertex has three legs whose secondary neighbors cannot form a consecutive chain.

By this definition, the 4-star is the only geographically good configuration that contains one inner vertex with four legs. Of the configurations whose inner vertices all have degree at least 5, those that are geographically good are precisely those that avoid the first two of Heesch's obstructions. In 1974, Appel and Haken achieved their objective; that is, they found an unavoidable set of geographically good configurations [APPEL and HAKEN 1976].

It would be advantageous to describe Appel's discharging algorithm in as much detail as the Dürre–Heesch algorithm. This, however, has already been done by Appel and Haken themselves in a most excellent way with an easy, accessible approach [APPEL and HAKEN 1976, pages 289–295]. Therefore, to do it here seems redundant. There is, however, a second reason for this omission. In this book, we are restricting ourselves to just making mention of the various discharging mechanisms that were used in [APPEL and HAKEN 1977]. They can be divided into two types. The first type consists of the discharging procedures used by Heesch and Mayer. In these, charges "flow" along edges from 5-vertices to neighboring major vertices. In fact:

1. With *regular* dischargings, or *R-dischargings*, the discharging value is 30.

2. With *small* dischargings, or *S-dischargings*, the discharging value is smaller than 30.

3. With *large* dischargings, or *L-dischargings*, the discharging value is greater than 30.

From their investigations, Haken and Appel noticed that the greatest difficulties were caused by the 6-6-chains, that is, edges that link two neighboring 6-vertices to one another. The second type, Haken's notion of a *transversal* discharging, dealt successfully with this problem. In the so-called *T-dischargings,* charges flow out from a 5-vertex transversally through several triangles and the 6-6-chains bounding them to a major vertex. The idea was good. However, Appel estimated that it would take over a year to convert this into a computer program. Therefore, together, they first attempted to produce a corresponding discharging procedure to be done by hand, and they succeeded in doing that. It is due to this success that discharging, as a component of the proof of the Four-Color Theorem, is not as computer-dependent in the same way as is the proving of reducibility. This is the second reason why we are not going into the same detail in regard to the Appel and Haken computer program.

For an explicit description of a discharging procedure, one must specify the individual cases together with the respective procedural mechanisms pertaining to each. It is particularly difficult when different discharging situations occur simultaneously in the same configuration. However, we will not go further into such issues at this time. Whoever wants to know at this point a little bit more about it should first study [APPEL and HAKEN 1986] and then [APPEL and HAKEN 1989], which is the updated version of [APPEL and HAKEN 1977].

The improvements to the proof of the Four-Color Theorem presented by Robertson, Sanders, Seymour, and Thomas [ROBERTSON et al. 1997] mainly concern the discharging procedures. They confirm a conjecture of Heesch's that in proving unavoidability, a reducible configuration can be found in the secondary neighborhood of an "overcharged" vertex. Because the outer vertices of a secondary neighborhood form a ring (Theorem 6.1.9), they were able to avoid the problems that arose with configurations having self-intersections (in other words, "immersion" problems—see page 160). These were a major cause of complications in Appel and Haken's joint work. The unavoidable set constructed by Robertson and his colleagues has 633 configurations, as opposed to 1476 in Appel and Haken's case. In addition, their discharging method (derived from another clever idea of Mayer's [MAYER 1978]) uses only 32 discharging rules, instead of the 300 + of Appel and Haken. Their proof of unavoidability is easier to check because, in their own words, they replaced *"the mammoth hand-checking of Appel and Haken by another mammoth handcheckable proof, but this time written formally so that, if desired, it can be read and checked by a computer in a few minutes."* Moreover, they developed a quadratic algorithm to color planar graphs with four colors. This is an improvement over the quartic algorithm used by Appel and Haken. The work of Robertson, Sanders, Seymour, and Thomas can be viewed on the internet. The URL is http://www.math.gatech.edu/ ~thomas/FC/fourcolor.html. All the necessary programs and data for a computer check of the entire proof are contained at this web site.

Bibliography

Textbooks, monographs, and contributions to collected works are marked with a * . Abbreviations for journals are given primarily in accordance with the "Zentralblatt für Mathematik," for older journals in accordance with "Jahrbuch über die Fortschritte der Mathematik," and for journals in conformance with the "Mathematical Reviews."

This list is itself in alphabetical order and therefore is not a part of the index.

AIGNER, M.
*1984   Graphentheorie—Eine Entwicklung aus dem
         4-Farbenproblem, Stuttgart: B. G. Teubner

ALLAIRE, F.
*1976   A minimal 5-chromatic planar graph contains a 6-valent
         vertex, Proceedings of the Seventh Southeastern
         Conference on Combinatorics, Graph Theory, and
         Computing, Louisiana State University, Baton Rouge,
         February 9–12, 1976, 61–78, Baton Rouge: Louisiana State
         University

ALLAIRE, F. and SWART, E.R.

1978   A systematic approach to the determination of reducible configurations in the four-color conjecture, J. Combinatorial Theory Ser. B **25**, 339–361

APPEL, K. and HAKEN, W.

1976   The existence of unavoidable sets of geographically good configurations, Illinois J. Math. **20**, 218–297

1977   Every planar map is four colorable, Part I: Discharging, Illinois J. Math. **21**, 429–490, also in [APPEL and HAKEN 1989]

*1978   The Four-Color Problem, Mathematics Today, 153–180, edited by L.A. STEEN, New York/Heidelberg/Berlin: Springer-Verlag

1986   The Four Color proof suffices, Math. Intell. **8**, 10–20

*1989   Every Planar Map is Four Colorable, Providence (Rhode Island): AMS

APPEL, K., HAKEN, W., and KOCH, J.

1977   Every planar map is four colorable, Part II: Reducibility, Illinois J. Math. **21**, 491–567, also in [APPEL and HAKEN 1989]

APPEL, K., HAKEN, W., and MAYER, J.

1979   Triangulation à v_5-séparés dans le problème des quatre couleurs, J. Combinatorial Theory Ser. B **27**, 130–150

BAKER, H.A. and OLIVER, E.G.H.

*1967   Ericas in Southern Africa, Cape Town/Johannesburg: Purnell & Sons

BALL, W.W.R.

*1892   Mathematical Recreations and Essays

*1939   *revised by* H.S.M. COXETER *11th edition*, London/New York: Macmillan and Co.

1915   Augustus de Morgan, Math. Gaz. **8**, 43–45

BALTZER, R.

1885   Eine Erinnerung an Möbius und seinen Freund Weiske, Leipz. Ber. **37**, 1–6

BELLOT, H.H.

*1929 University College London 1826–1926, London: University of London Press

BERGE, C.

*1958 Théorie des graphes et ses applications, Paris: Dunod

BERNHART, A.

1947 Six rings in minimal five color maps, Am. J. Math. **69**, 391–412

1948 Another reducible edge configuration, Am. J. Math. **70**, 144–146

BERNHART, F.R.

1978 Irreducible Configurations and the Four Color Conjecture, Theory and Applications of Graphs, Proceedings, Michigan, May 11-5,1976, edited by Y. Alavi and D. R. Lick, New York/Heidelberg/Berlin: Springer-Verlag

BIGALKE, H.-G.

*1988 Heinrich Heesch: Kristallgeometrie, Parkettierungen, Vierfarbenforschung, Basel/Boston/Berlin: Birkhäuser Verlag

BIGGS, N.L.

1983 De Morgan on map colouring and the separation axiom, Arch. Hist. Exact Sci. **28**, 165–170

BIGGS, N.L., LLOYD, E.K., and WILSON, R.J.

*1976 Graph Theory 1736–1936, Oxford: Clarendon Press

BIRKHOFF, G.D.

1913 The Reducibility of Maps, Am. J. Math. **35**, 115–128

BODENDIEK, R.

*1985 editor: Graphen in Forschung und Unterricht, Festschrift K. Wagner, Bad Salzdetfurth: Barbara Franzbecker

*1989 *see* WAGNER and BODENDIEK

BOUCHER, M.
*1974   The University of the Cape of Good Hope and the University of South Africa, 1873–1946, in Archives Year Book for South African History, Thirty-Fifth Year, Vol. I, Pretoria: The Government Printer

CAUCHY, A.-L.
1813   Recherches sur les polyèdres
1813   Journal de l'École polytechnique **9**, 68–86, excerpts printed in [BIGGS, LLOYD, and WILSON 1976]

CAYLEY, A.
1878   Nature **18**, 294
1879   On the colouring of maps, Proc. R. Geogr. Soc. **1**, 259–261, printed in [BIGGS, LLOYD, and WILSON 1976]
1889   Scientific Worthies XXV. James Joseph Sylvester, Nature **40**, 217–219

CIGLER, J. and REICHEL, H.C.
*1987   Topologie—Eine Grundvorlesung, 2. Auflage, Mannheim/Wien/ Zürich: B.I.-Wissenschaftsverlag

COXETER, H.S.M.
*1939   see BALL
1957   Map-coloring problems, Scripta Mathematica **23**, 11–25
1959   The four-color map problem, 1840–1890, Mathematics Teacher **52**, 283–289

CUNDY, H.M. and ROLLET, A.P.
*1961   Mathematical Models, 2nd edition, Oxford: Clarendon Press

DE MORGAN, AUGUSTUS
1852   Letter to W. R. Hamilton of October 23, preserved in the archives of Trinity College, Dublin, Hamilton mss., letter 668, excerpts printed in [MAY 1965] and [BIGGS, LLOYD, and WILSON 1976]
1853   Letter to W. Whewell of December 9, preserved in the archives of Trinity College, Cambridge, Whewell Add. mss., a.202^{125}, excerpts printed in [BIGGS 1983]

1854   *Letter to R. L. Ellis of June 24*, preserved in the archives of Trinity College, Cambridge, *Whewell Add. mss., c.*67[111], excerpts printed in [BIGGS 1983]

1860   *Anonymous review of the book "The Philosophy of Discovery" by W. Whewell*, Athenaeum **1694**, 501–503

DE MORGAN, SOPHIA ELIZABETH

1872   Memoir of Augustus de Morgan with selections of his writings, London: Longmans, Green & Co.

DINGELDEY, F.

*1890 Topologische Studien über die aus ringförmig geschlossenen Bändern durch gewisse Schnitte erzeugbaren Gebilde, Leipzig: B.G. Teubner

DIRAC, G.A.

1963   Percy John Heawood, J. Lond. Math. Soc. **38**, 263–277

DO CARMO, M.P.

*1976 Differential Geometry of Curves and Surfaces, Englewood Cliffs, New Jersey: Prentice-Hall Inc.

DUDENEY, H.E.

*1917 Amusements in Mathematics, Edinburgh: Thomas Nelson & Sons

1958   *2nd edition*, New York: Dover Publications

DUFF, J.

1955   Prof. P.J. Heawood O.B.E., Nature **175**, 368

DUGUNDJI, J.

*1966 Topology, Boston: Allyn and Bacon Inc.

DÜRRE, K.

*1969 Untersuchungen an Mengen von Signierungen (*Ph.D. thesis*), u. t. t.: Properties of sets of Colorations, Brookhaven: Associated Universities Inc.

DÜRRE, K. and MIEHE, F.

*1979 Eine Implementierung des Heesch-Algorithmus zur chromatischen Reduktion, Graphs, Data Structures, Algorithms, published by M. Nagl and H.-J. Schneider, München/Wien: Carl Hanser Verlag

EISELE, C.

*1976 *Introductions to the whole edition and the single volumes of* [PEIRCE 1976]

ELLIS, R. L.

*1863 The Mathematical and other Writings, published by WILLIAM WALTON, including a biographical sketch by HARVEY GOODWIN, Cambridge: Deighton, Bell and Co., and London: Bell and Daldy

ERRERA, A.

*1921 Du Coloriage de Cartes (*Ph.D. thesis*), Brüssel: Falk fils, Van Campenhout, successeur, and Paris: Gauthier–Villars

1925 Une contribution au problème des quatre couleurs, Bull. Soc. Math. Fr. **53**, 42–55

1927 Exposé historique du problème des quatre couleurs, Period. Mat. Serie 4, **7**, 20–41

EUCLID

*1883 Elementa, Buch I–IV *Greek and Latin*, published by I.L. Hieberg, Leipzig: Teubner

*1956 The thirteen books of Euclid's Elements translated from the text of Hieberg by T.L. Heath, Vol. I, 2nd ed., New York: Dover Publications, Inc.

EULER, L.

1750 *Letter to Christian Goldbach*[1]*November of 3/14 1750*, in [JUŠKEVIČ and WINTER 1965], excerpts printed in [BIGGS, LLOYD, and WILSON 1976]

1758a Elementa doctrinae solidorum, Novi commentarii academiae scientiarum Petropolitanae **4** (1752/53), 109–140, printed in [SPEISER 1953]

1758b Demonstratio nonnullarum insignium proprietatum quibus solida hedris planis inclusa sunt praedita, Novi commentarii academiae scientiarum Petropolitanae **4** (1752/53), 140–160, printed in [SPEISER 1953]

---

[1]Courtesy of the letter writer, next to the addressee is written the date—given not only in terms of the Gregorian calendar but also according to the Julian calendar, that in many places was still in current use.

FÁRY, I.

1947 On straight line representation of planar graphs, Acta Sc. Math. (Szeged) **11**, 229–233

FEDERICO, P. J.

*1982 DESCARTES on Polyhedra. A Study of the De Solidorum Elementis, New York/Heidelberg/Berlin: Springer-Verlag

FORSYTH, A. R.

1895 Obituary notices; Cayley, Proc. R. Soc. Lond. **58**, i–xviii

FRANKLIN, P.

1922 The Four Color Problem (*Ph.D. thesis*), Am. J. Math. **44**, 225–236

1938 Note on the Four Color Problem, J. Math. and Phys. **16**, 172–184

FRIDY, J.A.

1987 Introductory Analysis: The Theory of Calculus, New York: Harcourt Brace Jovanovich Inc.

FRITSCH, G. and R.

1991 Augustus de Morgan (1806–1871), Didaktik der Mathematik **19**, 247–251

FRITSCH, R.

1990 Wie wird der Vierfarbensatz bewiesen. Der mathematische und naturwissenschaftliche Unterricht **43**, 80-87

GELBAUM, B.R. and OLMSTED, J.M.H.

1964 Counterexamples in Analysis, San Francisco: Holden–Day Inc.

GRAVES, R.P.

1889 Life of Sir William Rowan Hamilton, Dublin: Hodges, Figgis, & Co.; London: Longmans, Green, & Co.

GUTHRIE, FREDERICK

1880 Note on the colouring of maps, Proc. R. Soc. Edinb. **10**, 727–728

HAKEN, W.G.R.

1973   An existence theorem for planar maps, J. Comb. Theory
       Ser. B **14**, 180–184

1976   *see* APPEL and HAKEN

HAMILTON, SIR W.R.

1857   Account of the icosian calculus, Proc. R. Ir. Soc. **6**, 415–416

HARARY, F.

1969   Graph Theory, Reading: Addison-Wesley Publishing
       Company

HEAWOOD, P.J.

1890   Map-colour theorem, Q.J. Math. Oxf. **24**, 332–338,
       excerpts printed in [BIGGS, LLOYD, and WILSON 1976]

1897   On the Four-colour Map Theorem, Q.J. Math. Oxf. **29**,
       270–285.

HEESCH, H.

1930   Zur systematischen Strukturtheorie III, IV (Über
       die vierdimensionalen Symmetriegruppen des
       dreidimensionalen Raumes, Über die Symmetrien zweiter
       Art in Kontinuen und Semidiskontinuen), Z. Krist. **73**,
       325–345, 346–356

*1969 Untersuchungen zum Vierfarbenproblem,
       Mannheim/Wien/Zürich: Bibliographisches Institut

1974   E–Reduktion, Preprint Nr. 19 des Instituts für Mathematik
       der Technischen Universität Hannover, printed in
       [BODENDIEK 1985]

HEFFTER, L.

1891   Über das Problem der Nachbargebiete, Math. Ann. **38**,
       477–508, excerpts printed in [BIGGS, LLOYD, and WILSON
       1976]

JUŠKEVIČ, A.P. and WINTER, E.

*1965 Leonhard Euler und Christian Goldbach: Briefwechsel
       1729–1764, Berlin: Akademie–Verlag and

*1978 Köln: Aulis Verlag Deubner & Co KG

KAINEN, P. C.

*1977   *see* SAATY and KAINEN

KEMPE, A. B.

1879   *Announcement in the section NOTES*, Nature **20**, 275

1879a   On the geographical problem of the four colors, Am. J. Math. **2**, 193–200, excerpts printed in [BIGGS, LLOYD, and WILSON 1976]

1879b   *Reference to the preceding paper*, Nation **756**, 440

1879c   How to colour a map with four colours, Nature **21**, 399–400

1879d   *Note without title in the APPENDIX*, Proc. Lond. Math. Soc. **10**, 229–231

1890   *Note without title in the APPENDIX*, Proc. Lond. Math. Soc. **21**, 456

1891   *Report of the meeting of the London Mathematical Society, April 9, 1891*, Proc. Lond. Math. Soc. **22**, 263

1922   *Obituary of* KEMPE *by Sir Charles S. Sherrington*, Proc. R. Soc. Lond. **102**, 375–376

KNOTT, C.G.

*1911   Life and Scientific Work of Peter Guthrie Tait, Cambridge: University Press

KOCH, J.

1976   Computation of Four Color Irreducibility, University of Illinois, Department of Computer Science technical report (UIUCDCS-R-76-802)

1977   *see* APPEL, HAKEN, and KOCH

KURATOWSKI, K.

1930   Sur les problèmes des courbes gauches en topologie, Fundam. Math. **15**, 271–283

LEBESGUE, H.

1940   Quelques consequences simple de la formule d'Euler, J. Math. Pures Appl. **9**, 27–43

LLOYD, E.K.

1976   *see* BIGGS, LLOYD, and WILSON

LONDON MATHEMATICAL SOCIETY

1878   *Report of the meeting of June 13, 1878*, Nature **18**, 294

MACMAHON, P.A.

1897   James Joseph Sylvester, Nature **55**, 492–494

MADDISON, I.

1897   Note on the history of the map-coloring problem, Bull. Am. Math. Soc. New Ser. **3**, 257

MACKENZIE, D.

1997   Slaying the kraken, Social Studies of Science, to appear 1998

MAXWELL, J.C.

1864   On reciprocal figures and diagrams of forces, Phil. Mag. IV. Ser. **27**, 250–261, printed in *Scientific Papers* **1**, 514–525

1869   On reciprocal figures, frames and diagrams of forces, Trans. R. Soc. Edinb. **26**, 1–40, printed in *Scientific Papers* **2**, 161–207

MAY, K.O.

1965   The origin of the Four–color Conjecture, Isis **56**, 346–348

MAYER, J.

1969   Le problème des régions voisines sur les surfaces closes orientables, J. Comb. Theory Ser. B **6**, 177–195

1975   Inégalités nouvelles dans le problème des quatre couleurs, J. Comb. Theory Ser. B **19**, 119–149

*1978   Une propriété des graphes minimaux dans le problème des quatre couleurs, Problèmes Combinatoires et Théorie des Graphes, Colloques internationaux C.N.R.S. No. 260, Paris: C.N.R.S.

1979   *see* APPEL, HAKEN, and MAYER

1980a   Une page mathématique de Valéry: le problème du coloriage des cartes, Bull. Études Valéryennes **25**, 31–43

1980b   Paul Valéry et le problème des quatres coulers, Regards
         sur la Théorie des Graphes, Actes du Colloque de
         Cerisy, 12–18 juin 1980, 263–267, Presses polytechniques
         Romandes

MENASCO, W.W. and THISTLETHWAITE, M.B.

1991   The Tait flyping conjecture, Bull. Am. Math. Soc. New
         Ser. **25**, 403–412

MIDONICK, H.O. (publisher)

1965   The Treasury of Mathematics, New York: Philosophical
         Library

MILNOR, JOHN

1976   Hilbert's Problem 18: On crystallographic groups,
         fundamental domains, and on sphere packing;
         Mathematical developments arising from Hilbert
         Problems (Proceedings of Symposia in Pure Mathematics
         Volume XXVIII); 491–506 (in particular, §2, pages
         498–499); Providence, Rhode Island: American
         Mathematical Society

MOISE, E.E.

*1977   Geometric Topology in Dimensions 2 and 3, New
         York/Heidelberg/Berlin: Springer-Verlag

MORSE, M.

1946   George David Birkhoff and his Mathematical Work, Bull.
         Am. Math. Soc. **52**, 357–391

NEIDHARDT, W.

1990   Monster-Kurven, Didaktik der Mathematik **18**, 183–209

OLIVER, E.G.H.

*1967   see BAKER and OLIVER

OLMSTED, J.M.H.

*1964   see GELBAUM and OLMSTED

ORE, O.

*1967   The Four Color Problem, New York: Academic Press

ORE, O. and STEMPLE, J.G.

1970   Numerical calculations on the Four-color Problem, J. Comb. Theory Ser. B **8**, 65–78

OSGOOD, T.W.

*1973   An existence theorem for planar triangulations with vertices of degree five, six and eight, (*Ph.D. thesis*), University of Illinois

OSGOOD, W.F.

1903   A Jordan curve of positive area, Trans. Am. Math. Soc. **4**, 107–112

PEIRCE, C.S.

1880a   *see* SCIENTIFIC ASSOCIATION at JOHNS HOPKINS UNIVERSITY

*1976   The New Elements of Mathematics, *unpublished manuscript*, publisher C. EISELE, Den Haag/Paris: Mouton Publishers and Atlantic Highlands (New Jersey): Humanities Press

RATIB, I. and WINN, C.E.

1936   Généralisation d'une réduction d'Errera dans le Problème des Quatre Couleurs, C.R. Congr. Int. Math. Oslo 1936, 131, Oslo: A.W. Brøggers Boktrykkeri A/S

REYNOLDS Jr., C.N.

1926   On the problem of coloring maps in four colors, I., Ann. Math. **28**, 1–15

1927   On the problem of coloring maps in four colors, II., Ann. Math. **28**, 477–492

RINGEL, G.

*1959   Färbungsprobleme auf Flächen und Graphen, Berlin: Deutscher Verlag der Wissenschaften

*1974   Map Color Theorem, Berlin/Heidelberg/New York: Springer–Verlag

RINGEL, G. and YOUNGS, J.W.T.

1968 Solution of the Heawood map–coloring problem, Proc. Natl. Acad. Sci. USA **60**, 438–445

RINOW, W.

*1975 Lehrbuch der Topologie, Berlin: Deutscher Verlag der Wissenschaften

RITCHIE, W.

1918 The History of the South African College 1829–1918, vol.1, Capetown: T. Maskew Miller

ROBERTSON, N., SANDERS, D.P., SEYMOUR, P.D., and THOMAS, R.

1996 A new proof of the four-colour theorem, Electronic Research Announcements of the American Mathematical Society 2, no. 1, 17–25 (electronic)

1997 The Four-Colour Theorem, J. Comb. Theory Ser. B **70**, 2–44

ROLLET, A.P.

1967 *see* CUNDY and ROLLET

SAATY, T.L.

1972 Thirteen colorful variations on Guthrie's Four–colour Conjecture, Amer. Math. Monthly **79**, 2–43

SAATY, T.L. and KAINEN, P.C.

1977 The Four-color Problem: Assaults and Conquest, New York: McGraw–Hill

SALMON, G.

1883 Science worthies XXII. Arthur Cayley, Nature **28**, 481–485

SANDERS, D.P.

1996 *see* ROBERTSON, SANDERS, SEYMOUR, and THOMAS

1997 see ROBERTSON, SANDERS, SEYMOUR, and THOMAS

SCHMIDT, E.

1923 Über den Jordanschen Kurvensatz, Berl. Ber. 318–329

SCHNEIDER, I.

1991   Die Geschichte von der Begegnung zwischen dem
        Rechenmeister und dem Philosophen—Besuchte
        Descartes den Rechenmeister Faulhaber im Winter
        1619/20? Kultur und Technik, Z. des Deutschen Museums,
        Heft 4, 46–53

1993   Johann Faulhaber—Rechenmeister in einer Welt des
        Umbruchs, Basel/Boston/Berlin: Birkhäuser Verlag

SCIENTIFIC ASSOCIATION at JOHNS HOPKINS UNIVERSITY

1880   *Report* from the November 5, 1879 meeting, Johns
        Hopkins University Circular **1**, 16

1880a   *Report* from the December 3, 1879 meeting, Johns
        Hopkins University Circular **1**, 16

SEYMOUR, P.D.

1995   Progress on the four-color theorem, pp. 183–195
        in: Proceedings of the International Congress of
        Mathematicians, ICM '94, August 3–11, 1994, Zürich,
        Switzerland. Vol. I, edited by S. D. CHATTERJI,
        Basel/Boston/Berlin: Birkhäuser Verlag

1996   *see* ROBERTSON, SANDERS, SEYMOUR, and THOMAS

SPEISER, A.

1953   Editor: *Leonhardi Euleri Opera Omnia,* Series prima, *Opera
        Mathematica,* Vol. **XXVI**. Commentationes Geometricae,
        Zürich

STANIK, R.

1973   Zur Reduktion von Triangulationen (*Ph.D thesis*),
        Hannover

STEMPLE, J.

1970   *see* ORE and STEMPLE

STORY, W.E.

1879   Note on the preceding paper [= KEMPE 1879a], Am. J.
        Math. **2**, 201–204

STROMQUIST, W.R.

1975a   Some aspects of the Four–color Problem, (*Ph.D. thesis*), Harvard University

1975b   The Four–color Theorem for Small Maps, J. Comb. Theory Ser. B **19**, 256–268

SWART, E. R.

1978   *see* ALLAIRE and SWART

TAIT, P.G.

1880   On the colouring of maps, Proc. R. Soc. Edinburgh **10**, 501–503

1880   Remarks on the previous Communication, ([GUTHRIE 1880]), Proc. R. Soc. Edinburgh **10**, 729, printed in [BIGGS, LLOYD, and WILSON 1976]

1880   Note on a theorem in the geometry of position, Trans. R. Soc. Edinb. **29**, 657–660, printed in *Scientific Papers* **1**, 408–411

1884   On Listing's topology, Phil. Mag. V. Ser. **17**, 30–46, printed in *Scientific Papers* **2**, 85–98

THISTLETHWAITE, M.B.

1991   *see* MENASCO and THISTLETHWAITE

THOMAS, R.

1996   *see* ROBERTSON, SANDERS, SEYMOUR, and THOMAS

TODHUNTER, J.

*1876   William Whewell D.D. Master of Trinity College, Cambridge. An account of his writings with selections from his literary and scientific correspondence, *2 volumes*, London: Macmillan

TOEPELL, M.

*1991a   Mitgliedergesamtverzeichnis der Deutschen Mathematiker-Vereinigung 1890–1990, München: Inst. Gesch. Naturw. Univ. München

TUTTE, W.T.

1946   On Hamiltonian circuits, J. Lond. Math. Soc. **21**, 98–101

1948   On the Four Colour Conjecture, Proc. Lond. Math. Soc. **50**, 137–149

1972   *see* WHITNEY, H.

1974   Map-coloring problems and chromatic polynomials, Am. Scientist **62**, 702–705

*1975   Chromials, Studies in Graph Theory II, 361–377, edited by D.R. Fulkerson: The Mathematical Association of America

1978   Colouring Problems, The Math. Intell. **1**, 72–75

VEBLEN, O.

1912   An application of modular equations in analysis situs, Ann. Math., Ser. 2 **14**, 86–94

1947   George David Birkhoff (1884–1944), Am. Phil. Soc., Year book 1946, 279–285

VON KOCH, H.

1904   Sur une courbe sans tangente, obtenue par une construction géométrique élémentaire, Ark. Mat. Astr. Fys. **1**, 681–702

WAGNER, K.

1936   Bemerkungen zum Vierfarbenproblem, Jahresber. Dtsch. Math-Ver. **46**, 16–32

*1970   Graphentheorie, Mannheim/Wien/Zürich: Bibliographisches Institut

WAGNER, K. and BODENDIEK, R.

*1989   Graphentheorie I—Anwendungen auf Topologie, Gruppentheorie und Verbandstheorie, Mannheim/Wien/Zürich: B.I. Wissenschaftsverlag

*1990   Graphentheorie II—Weitere Methoden, Masse-Graphen, Planarität und minimale Graphen, Mannheim/Wien/Zürich: B.I. Wissenschaftsverlag

WERNICKE, P.

1904   Über den kartographischen Vierfarbensatz, Math. Ann. **58**, 419

WHEWELL, W.

*1860   On the Philosophy of Discovery, Chapters Historical and Critical, including the completion of the third edition of the philosophy of the inductive sciences, London: John W. Parker and Son, West Strand

WHITNEY, H. and TUTTE, W.T.

1972   Kempe chains and the Four Colour Problem, Util. Math. **2**, 241–281

1975   printed in *Studies in Graph Theory, Part II*, 378–413, published by D.R. Fulkerson, Math. Ass. Am.

WILSON, R.J.

*1976   *see* BIGGS, LLOYD, and WILSON

WILSON, J.

1976   New light on the origin of the Four-color Conjecture, Hist. Math. **3**, 329–330

WINN, C.E.

1936   *see* RATIB and WINN

1937   A class of coloration in the Four Color Problem, Am. J. Math. **59**, 515–528

1938   On certain reductions in the Four Color Problem, J. Math. and Phys. **16**, 159–171

1940   On the minimal number of polygons in an irreducible map, Am. J. Math. **62**, 406–416

WINTER, E.

*1965   *see* JUŠKEVIČ and WINTER

YOUNGS, J.W.T.

1968   *see* RINGEL and YOUNGS

Works of Reference

Biographisch-Literarisches Handwörterbuch zur Geschichte der Exakten Wissenschaften, Bde. 1-7b published by J.C. Poggendorf, Leipzig/Berlin: 1863/1990

Bell, E.T., Die großen Mathematiker Düsseldorf/Wien: 1967 Econ–Verlag

Chambers's Encyclopædia 15 volumes. London: 1959/1964

Concise Dictionary of American Biography New York: [3]1980

The New Century Cyclopedia of Names 3 volumes. New York: 1954

Dictionary of American Biography 22 volumes. Oxford: 1928/1958

The Dictionary of National Biography 63 volumes. London/Oxford: 1885/1990

Dictionary of Scientific Biography 14 volumes. New York: 1976 (Index 1980)

Dictionary of South African Biography
5 volumes. Cape Town/Pretoria: 1976/1987

The Encyclopedia Americana
30 volumes. Danbury (Connecticut): 1986

The New Encyclopædia Britannica Micropædia
12 volumes. Chicago: [15]1985

Englisches Reallexikon
2 volumes. published by C. KLÖPPER, Leipzig: 1897

Index Herbariorum, Part I: The Herbaria of the World
Editors: P. K. HOLMGREN, N. H. HOLMGREN, and L. C. BARNETT,
Bronx: 1990

Who Was Who
London: 1929

Wußing, H. und Arnold, W. Biographien bedeutender Mathe-
matiker
Berlin: 1975 Volk und Wissen Volkseigener Verlag
Köln: 1978 Aulis Verlag Deubner & Co KG

Index